AutoCAD Release 13

A Supplement for Engineering Design Graphics

JAMES H. EARLE

Texas A & M University

An imprint of Addison Wesley Longman, Inc.

Menlo Park, California • Reading, Massachusetts • Harlow, England
Berkeley, California • Don Mills, Ontario • Sydney • Bonn • Amsterdam • Tokyo • Mexico City

Senior Production Editor: Teri Hyde
Composition and Film Buyer: Vivian McDougal
Manufacturing Supervisor: Janet Weaver
Cover Design: Frank Hom
Composition: Fog Press

Access the latest information about Addison-Wesley books from our World Wide Web page:
http://www.aw.com

Printed in the United States of America.

ISBN 0–201–17997–0

1 2 3 4 5 6 7 8 9 10 CRS 00 99 98 97 96

Contents

Preface

This supplement has been designed for you to use with *Engineering Design Graphics, Eighth Edition* as an introduction to AutoCAD® Release 13. Chapters One and Two of the supplement are created to be parallel with Chapters 36 and 37 of *Engineering Design Graphics* so you can make an easy transition from AutoCAD Release 12 to AutoCAD Release 13. If you are already using Release 13 in Windows, you can easily substitute these chapters for their parallel chapters within the text.

Some of the new features of AutoCAD Release 13 covered in this supplement are:

- the AutoCAD for Windows Drawing Screen
- Flyouts
- Floating Toolbars
- Tooltips
- the new Menu System
- Rendering Enhancements
- Associative Hatching
- Boundary Hatching
- Dimensioning Styles
- Layer and Linetype Display
- Solid Primitives

We are pleased to offer additional resources for the study and teaching of AutoCAD on our ftp sight. Connect to *ftp.aw.com/cseng/authors/lockhart/r13.* If you have trouble, connect to *ftp.aw.com* and change to the *cseng/authors/lockhart/r13 directory.* Use anonymous as your user name and your email address as your password when you log on. You can download an AutoCAD Release 13 Command Reference, a Disk Safety Cartoon, transparency masters showing screen shots of AutoCAD Release 13, and other helpful tools from this sight.

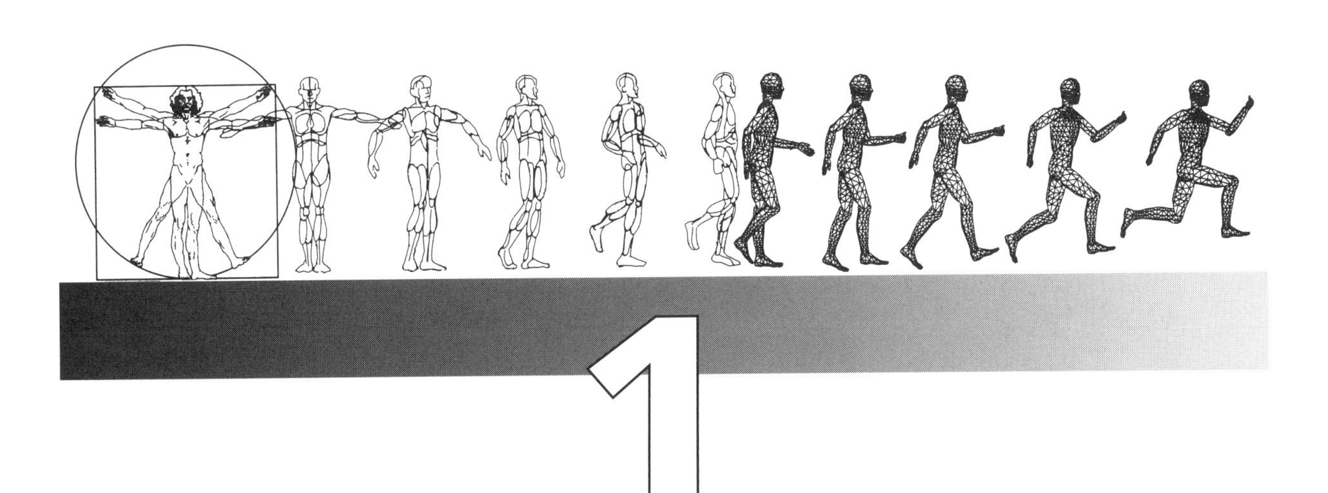

2D Drawing, Dimensioning, and Sketching with AutoCAD Release 13

1.1 Introduction

This chapter provides an introduction to computer graphics by using AutoCAD, Release 13, which runs on a 486 or faster computer, with at least 16 Mb of RAM and a 100-Mb hard disk, a mouse (or tablet), and an A-B plotter. AutoCAD was selected as the software for presenting computer graphics because it is the most widely used.

The coverage of AutoCAD in this supplement is brief and many operations have been omitted entirely because of space limitations. AutoCAD's concisely written *User's Guide* and *Command Reference* manuals take up 87,890 Kb on the CD-ROM. However, AutoCAD is covered here sufficiently to guide you through applications typical of an engineering design graphics course.

1.2 An Experimental Session

If this is your first session, you are anxious to turn the computer on, make a drawing on the screen, and plot it without having to read instructions. This section is what you're looking for.

Starting AutoCAD Load AutoCAD by double-clicking on the `AutoCAD R13` icon to obtain the drawing screen **(Fig. 1.1)**. Move the cursor around the screen with your mouse, select items, and try the pull-down menus from the menu bar.

Mouse Most interactions with the computer will be accomplished with a **mouse (Fig. 1.2),** but most commands can be entered at the keyboard more quickly after you learn them. Press the left button to click on, select, or pick a command or object; a "double click" is needed in some cases.

Creating a File To create a new file, pick `File` from the menu bar and select `New...` from the pull-down menu **(Fig. 1.3)**. Then, type the drawing name, `DRW-1`, in the edit box of the `Create New Drawing` dialog box **(Fig. 1.4)**.

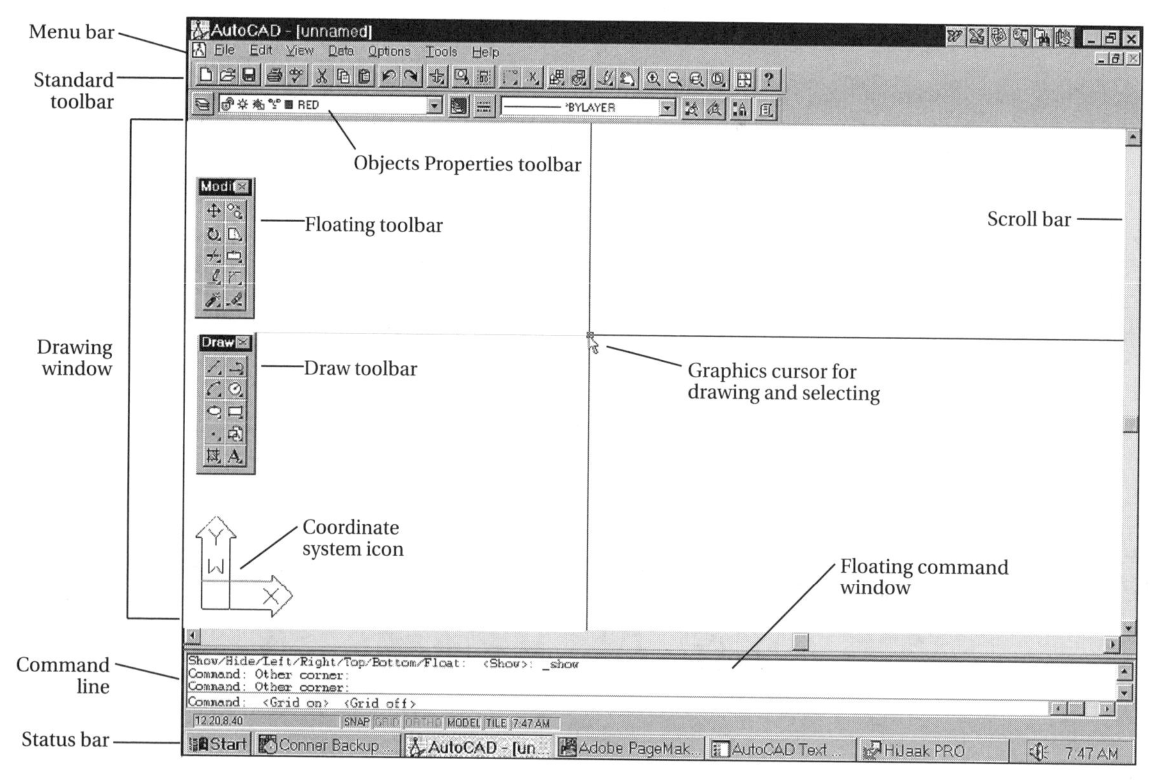

Figure 1.1 After loading AutoCAD, the screen will appear ready for drawing.

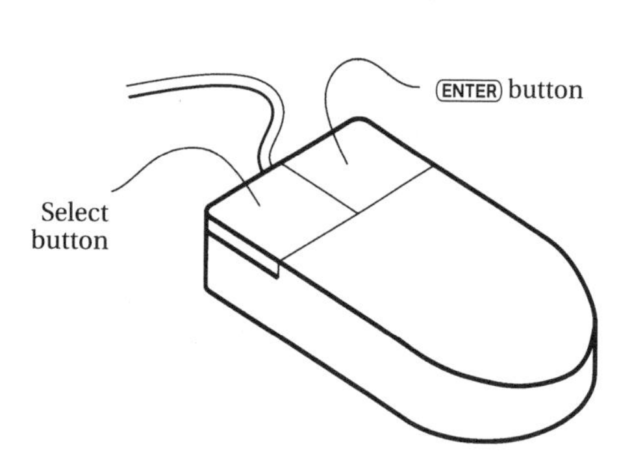

Figure 1.2 The left button on the mouse is the select button, and the right button is the ENTER button.

FILE: NEW

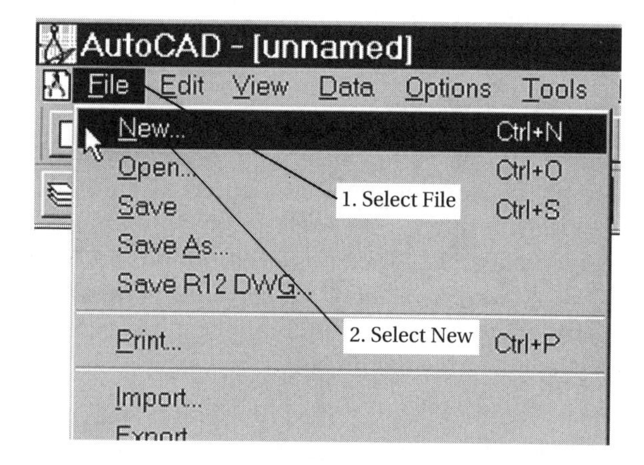

Figure 1.3 To begin the procedure for creating a new file, select `File` and `New`.

CREATE NEW DRAWING

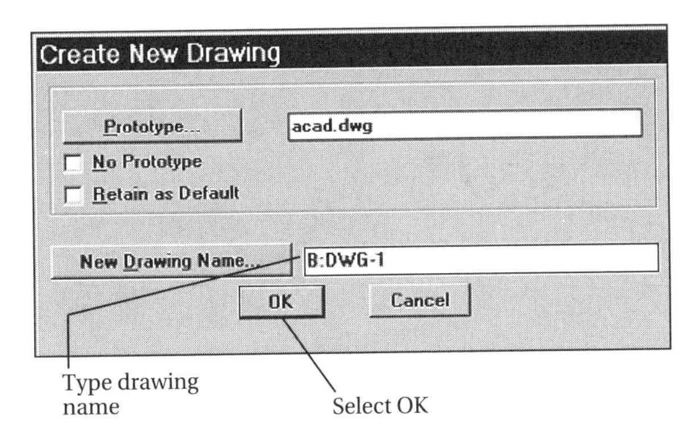

Figure 1.4 The `Create New Drawing` dialog box lets you assign a prototype file (`acad.dwg`) to a new file (`DRW-1`) and press `OK` to proceed.

FILE: SAVE

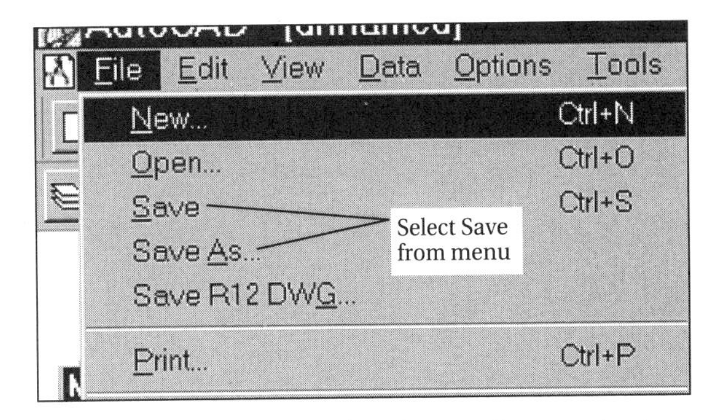

Figure 1.6 Select `Save` under `File` on the menu bar to save your drawing on a disk.

DRAWING ON THE SCREEN

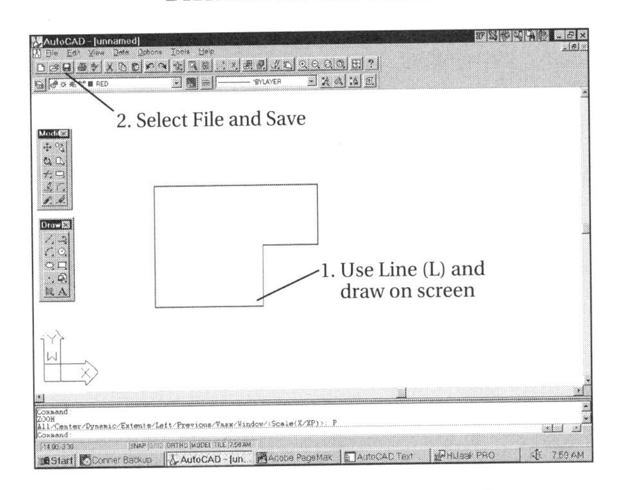

Figure 1.5 By typing `L` or `Line`, or selecting the `LINE` icon, lines can be drawn on the screen by selecting points with the mouse.

Making a Drawing Type `L`, press (ENTER) and draw some lines on the screen with the mouse for the fun of it as shown in **Fig. 1.5**. Move the **cursor** (the crosshairs controlled by the mouse) to the `Draw` toolbar and pick the `LINE` icon to get the prompt `line From point:`, in the `Command` line at the bottom of the screen. Select an endpoint with the mouse by clicking the left button, move the cursor to a second point, and other points. To disconnect the rubber band from the current point, press the right button of your mouse, which is the same as the (ENTER) key. Try drawing a circle and other objects on the screen by selecting icons from the `Draw` toolbar.

Repeat Commands By pressing (ENTER) after the previous command, the command can be repeated. For example, `LINE` will appear in the Command line at the bottom of the screen after pressing (ENTER), if `LINE` was the previous command.

Updating Your File To save the file `DRW-1`, click on `File` from the menu bar and `Save` from the pull-down menu **(Fig. 1.6)**. The command `_qSave` appears in the `Command` line at the bottom of the screen and `DRW-1` is updated.

Plotting Your Drawing Select `File` and `Print` from the menu bar **(Fig. 1.7)** and the `Plot Configuration` dialog box will be displayed

FILE: PRINT

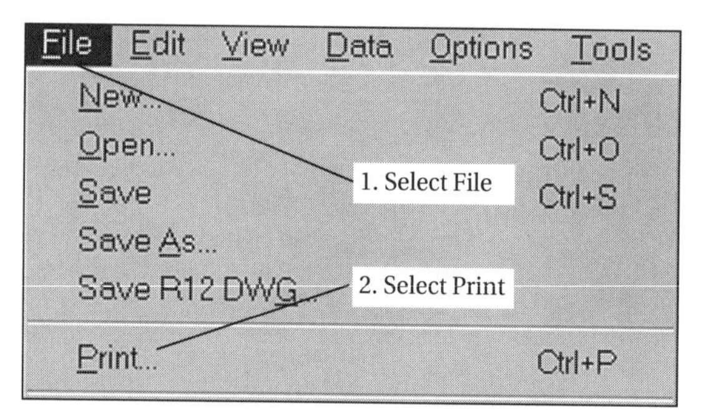

Figure 1.7 To prepare to plot a drawing, select `Print`, from the `File` option of the menu bar.

PLOT CONFIGURATION

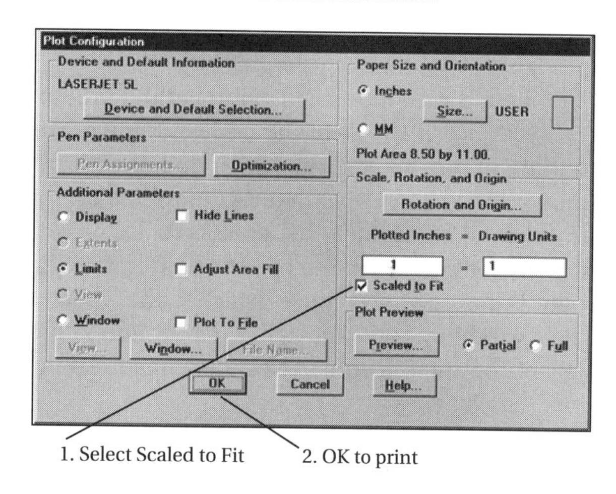

Figure 1.8 The `Plot Configuration` dialog box will appear on the screen for plotting instructions. Select `Extents`, `Scaled to Fit`, and `OK` to plot.

(Fig. 1.8). Select the `Extents` and `Scaled to Fit` buttons so the plotted drawing will fill the sheet. Load the A-size paper sheet in the plotter, which may be a Hewlett-Packard 7475 that plots both A-size and B-size sheets with ink pens. Select the `OK` button and the drawing data are sent to the plotter.

Ending the Session Select `File` and `Exit` from the menu bar **(Fig. 1.9)** to end the session. When

EXITING AUTOCAD

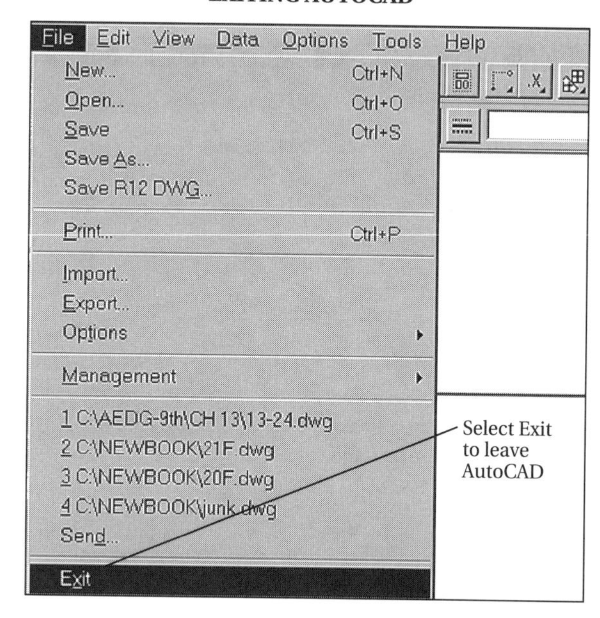

Figure 1.9 Select `Exit` from the `File` pull-down menu to exit AutoCAD.

working from a floppy disk, do not remove it from its drive until it has been saved with either the `Save` or `Save As` options.

That's how it works. Now, let's get into the details.

1.3 Introduction to Windows

Although Release 13 can be used with DOS, Windows is the recommended operating system. The **screen menu** can be activated by selecting `Options` on the menu bar, `Preferences` from the pull-down menu, and the `Screen Menu` check box in the `Preferences` dialog box **(Fig. 1.10)**. The screen menu will appear at the right side of the screen as shown in **Fig. 1.11**.

By pressing the **escape key** (ESC), the current command is aborted **(Fig. 1.12)**. The word `Command:` appears in the `Command` line at the

PREFERENCES BOX (OPTIONS TOOLBAR)

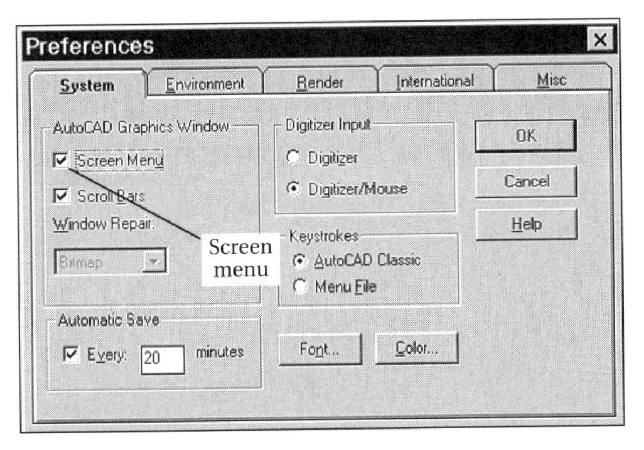

Figure 1.10 Select `Screen Menu` from the `Preferences` box (under `OPTIONS`) to obtain the screen menu on the screen as an alternate method of selecting commands.

bottom of the screen, ready for a new command. The (ESC) key is used to abort a plot operation, but there will be a delay until the data have been emptied from the buffer.

Windows allows several programs to be open and running at the same time. To move from one program to another, hold the (ALT) key down and press the (TAB) key several times until the icon of the desired program appears, then release the buttons **(Fig. 1.13)**. To close a window, double click on the minus sign (–) in the upper left border. Remember to use the `HELP` command whenever additional instructions are needed.

AUTOCAD FOR WINDOWS SCREEN

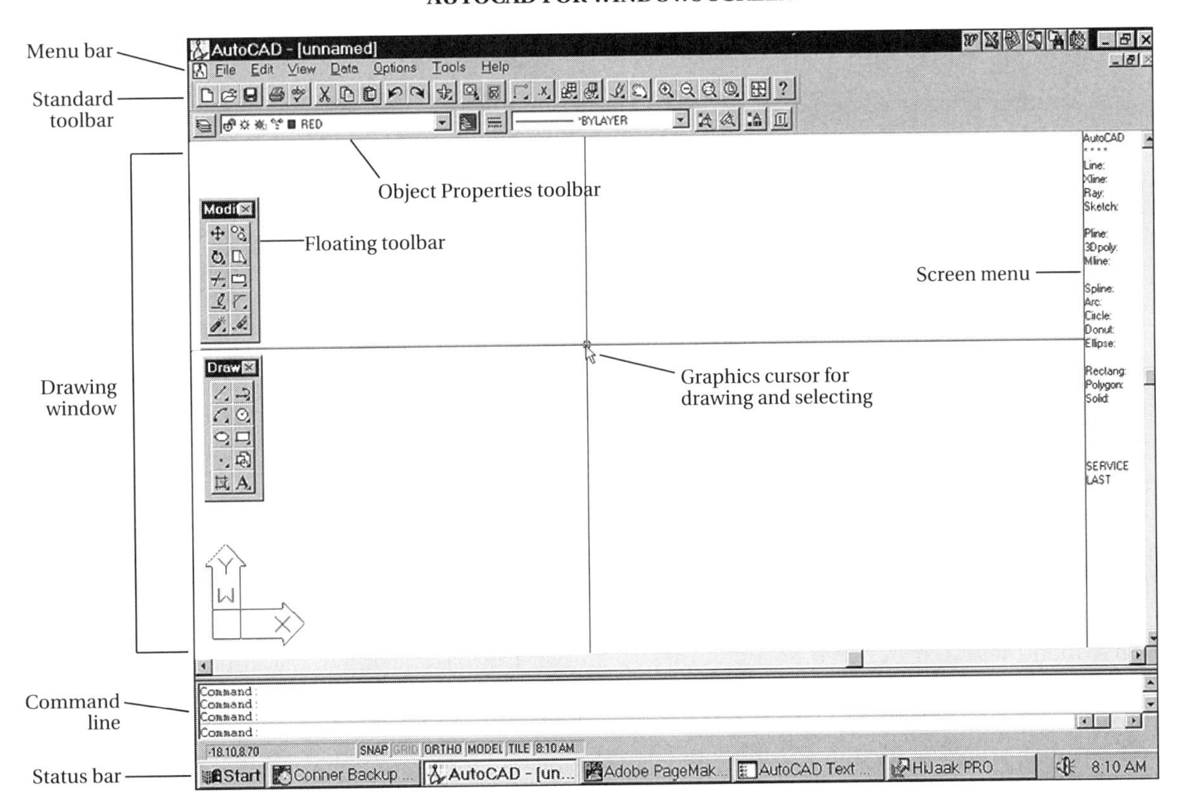

Figure 1.11 The typical screen is shown here, ready for drawing.

ABORTING A COMMAND

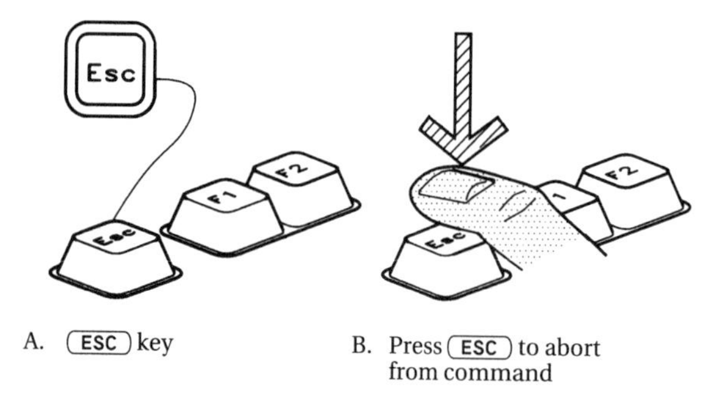

Esc
A. ESC key
B. Press ESC to abort from command

Figure 1.12 The current operation is aborted by pressing the escape key ESC.

OPENING OTHER PROGRAMS

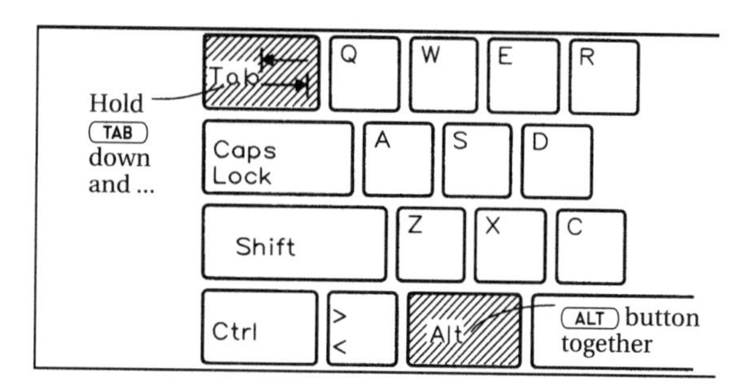

Figure 1.13 To cycle sequentially through the programs that are open in `Windows`, hold down the ALT key and repetitively press the TAB key.

1.4 Using Dialog Boxes and Toolbars

AutoCAD Release 13 has many **dialog boxes** with command names beginning with `DD` (`DDLMODES`, for example) to interact with the user. The command `FILEDIA` can be used to turn off (0 = on and 1 = off) the dialog boxes if you prefer to type the commands without dialog boxes. When a command on a menu followed by three dots (`...`) is selected, additional dialog boxes will be displayed. Some dialog boxes have **subdialog boxes**.

Tooltips are provided to identify the functions of each tool on a toolbar. By placing the cursor on a tool icon, a tooltip will appear to define its function, as shown in **Fig. 1.15.**

LOADING A TOOLBAR

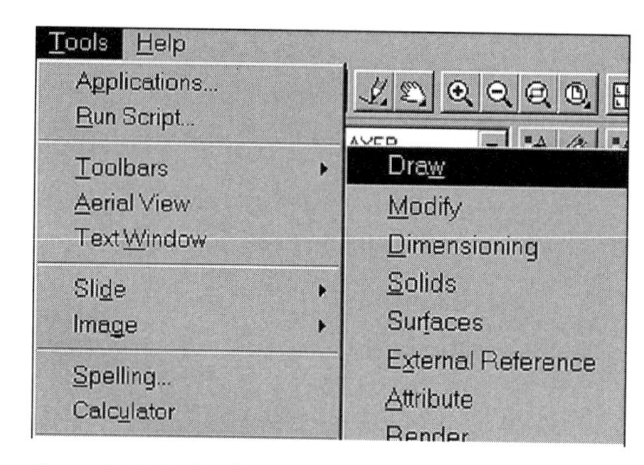

Figure 1.14 To load a `toolbar`, select `Tools`, `Toolbars`, and `Draw`, which is the name of this particular toolbar.

TOOLBARS, FLYOUTS, AND TOOLTIPS

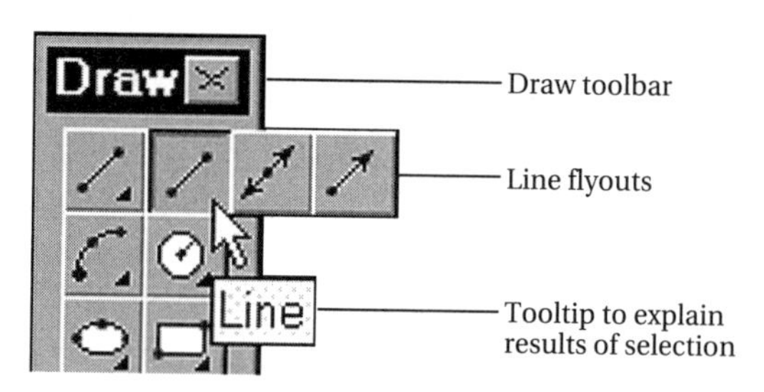

Figure 1.15 Many commands selected from the `toolbars` have `flyouts` for multiple options. By placing the pointer on a box, its definition is given in a tooltip.

Double clicking the mouse (quickly pressing the select button twice) selects a file from a file list. Double clicking is also used as an alternative to making the selection followed by `OK`. By experimentation you will soon learn where double clicking is applied.

The `Select File` dialog box in **Fig. 1.16** has

OPENING A DRAWING

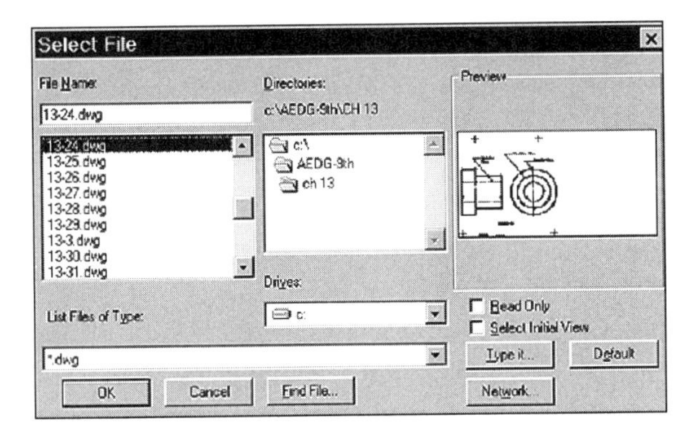

Figure 1.16 Under the `File` heading, select `Open` to obtain the `Select File` dialog box that will display the file when its name is selected from the list.

SCROLLING BARS

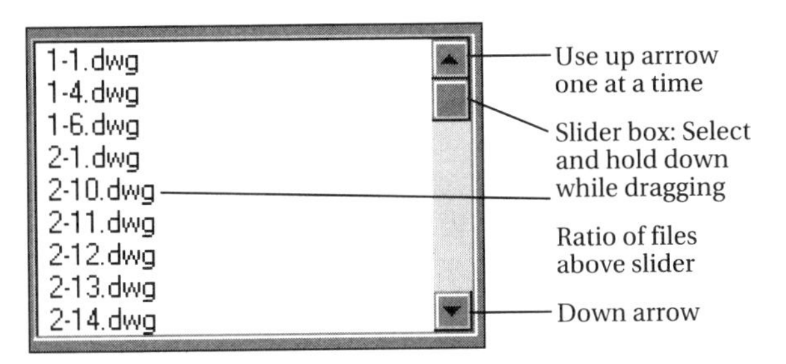

Figure 1.17 Scroll through a list of files by using scroll bars, and select the slide box while pressing the select button of the mouse. Directional arrows at each end can be used to move up or down the list one at a time.

lists, blanks, and buttons that can be selected by the cursor. When a file is selected, it is darkened by a gray bar and a thumbnail of it is shown in the window.

Check boxes are used to toggle options `On` (an `X` is shown) or `Off` (the box is blank). **Edit boxes** are boxes in which to type responses such as the name of the file to be saved (Fig. 1.16). **Radio buttons** are circles with circles inside them that can be turned on or off.

Scroll bars are used to move through lists in some dialog boxes via the up and down arrows **(Fig. 1.17)**. Flyouts appear when a command followed by a triangle is selected as shown in Fig. 1.15. After selecting `LINE`, you must define its type in a flyout. The `OK` button is used to activate the selections made in the dialog box.

Your speed will increase if you learn and type commands at the `Command` line instead of using dialog boxes. What could be easier or faster than typing `L` for `LINE` or `C` for `CIRCLE`?

1.5 Drawing Aids

Type `LIMITS` to establish the size of the drawing area that will be filled with dots when `Grid` is typed and set `On` (or select `Drawing Limits` from the `Data` menu and set `On`). A drawing that fills an A-size sheet (11 × 8.5 inches) has a plotting area of about 10.1 × 7.8 inches or 257 × 198 mm. The `Drawing Limits` command is found under the `Data` menu of the menu bar, the screen menu, and it can be typed at the Command line as follows:

`Command:` <u>`LIMITS`</u> (ENTER)

`ON/OFF/<Lower left corner> <0.00,0,00>:` (ENTER) `(Accept default value.)`

`Upper right corner <12.00,9.00>:` <u>`11,8.5`</u> (ENTER)

`Limits` can be reset at any time during the drawing session by repeating these steps.

`Units` sets the format of numerals and their fractional parts. Under the `Data` menu, select `Units` and the `Units Control` dialog box (`DDUNITS` command) appears **(Fig. 1.18)**. `Units` formats are set by selecting one of the following options:

1. Scientific 1.55E+01
2. Decimal 15.50
3. Engineering 1'–3.50"
4. Architectural 1'–3$\frac{1}{2}$"
5. Fractional 15$\frac{1}{2}$

UNITS CONTROL BOX

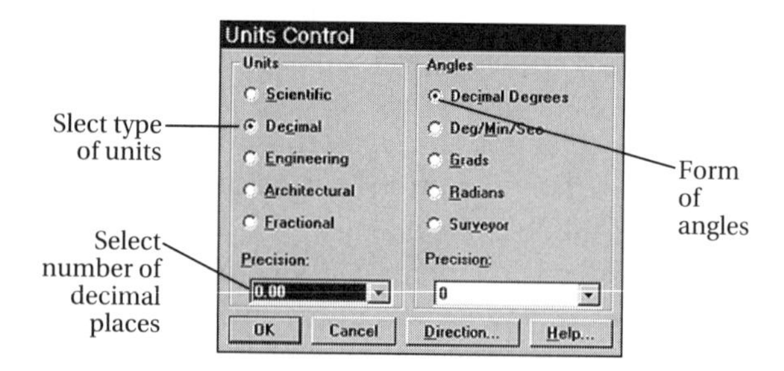

Figure 1.18 The `Units Control` dialog box, under `Data` of the menu bar, lets you set the number of decimal places and the form of numbers and angles.

DRAWING AIDS BOX (DDRMODES)

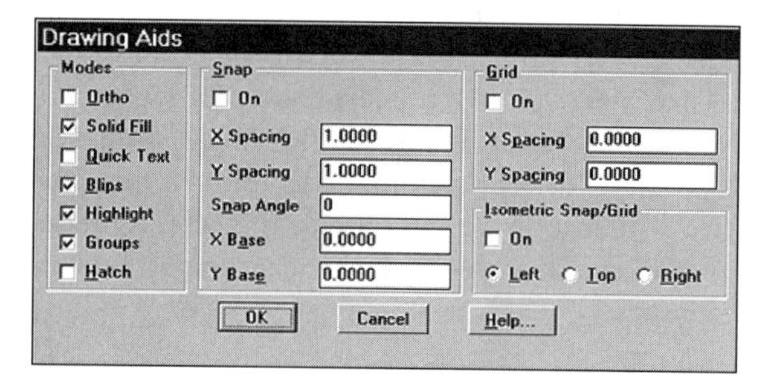

Figure 1.19 The `Drawing Aids` dialog box is used to set `Modes`, `Snap`, and `Grid` settings.

The number of decimal places for fractions is obtained by picking the `Precision` edit box for a list of options. `Angular Units` are selected in the same manner.

The `Drawing Aids` dialog box (`DDRMODES`) is found under the `Options` menu from which `Modes`, `Snap`, and `Grid` are available **(Fig. 1.19)**.

Function keys, shown in **Fig. 1.20** are used to toggle `On` and `Off` several of the drawing aids:

`Ortho` forces all lines to be either horizontal or vertical (not angular). `Ortho` can be set by picking the check box in Fig. 1.19, by pressing the (F6) key, or by typing `Ortho` and `On` or `Off`.

FUNCTION KEYS

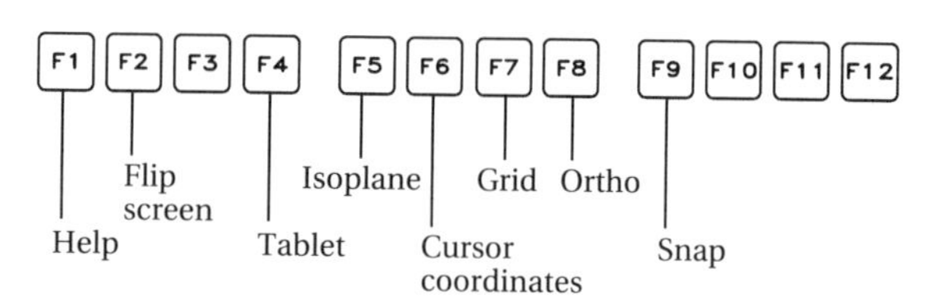

Figure 1.20 These function keys can be used to turn the settings indicated either `Off` or `On`.

When `On`, `Ortho` appears in the status line at the bottom of the screen.

`Solid Fill` can be turned on or off to show areas drawn with the `SOLID` command as filled or outlined areas.

`Quick Text` saves regeneration time by showing text as boxes. Text is restored when `QTEXT` is turned `Off` and `REGEN` is typed.

`BLIPMODE` is set `On` or `Off` by selecting `BLIPS` from the dialog box or by typing `BLIPMODE` and `On` or `Off`. `Blips` are temporary markers made on the screen when selections are made; they are removed by refreshing the screen (pressing (F7)).

`SNAP` forces the cursor to stop at points on an imaginary grid of a specified spacing. The X- and Y-spacings of `SNAP` can be typed in the edit boxes of the `Drawing Aids` dialog box, or begun from the screen menu or the `Command` line as follows:

`Command:` <u>`SNAP`</u> (ENTER)

`Snap spacing or ON/OFF/Aspect/Rotate/Style<0.25>.`<u>`20`</u>

This response sets `SNAP` to 0.20 units. When `On`, `SNAP` appears in the status bar at the bottom of the screen. `SNAP` can be toggled on and off by pressing (F9) or double clicking on its `Status Bar` icon.

Select `ASPECT` by typing `A` to assign X- and Y-values if they are different. The `ROTATE` option

STATUS BAR (BOTTOM OF SCREEN)

Command: LINE
1.70,0.10 SNAP GRID ORTHO MODEL TILE 10:38 AM

Cursor x and y coordinates
SNAP: On or Off
GRID: On or Off
ORTHO: On or Off
Model Space
TILEMODE = 1
Time

Figure 1.21 The `Status bar` at the bottom of the screen displays the current settings. Double click on the Drawing Aid boxes to turn them on or off.

HELP SCREEN

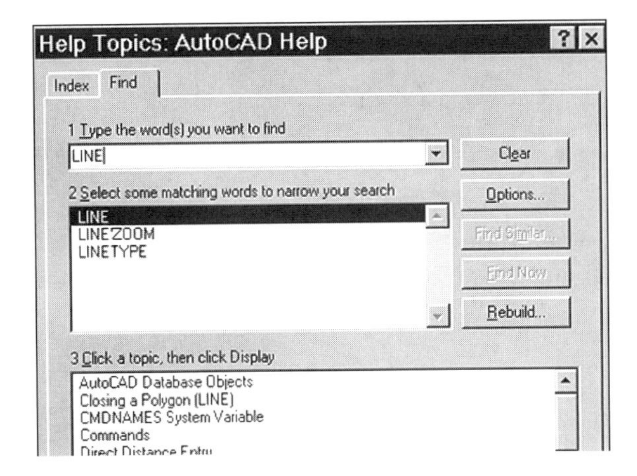

Figure 1.22 By selecting `HELP` from the menu bar, this `Search` box appears with a slider bar for scrolling through the topics that can be selected for help.

prompts for the angle of rotation of the `SNAP` grid and its origin point, which is given by typing X- and Y-values in the edit boxes.

`Snap Style` can be set to `I` (isometric) or `S` (the standard rectangular pattern). The `I` option is used for isometric drawings with axes 120° apart.

`GRID` (`F7`) fills the `LIMITS` area with a dot spacing assigned by typing values in the X- and Y-spacing boxes. By typing `GRID` and selecting the `SNAP` option, the grid is set equal to the `SNAP` spacing.

Running coordinates are displayed at the bottom of the screen in the status bar to show the cursor's position by pressing (F6).

The `Status Bar` at the bottom of the screen displays several of the settings discussed above as black if `On` or gray if `Off` **(Fig 1.21)**. Double clicking on these buttons toggles them off or on.

1.6 General Utility Commands

Utility commands control various operations to make changes in files.

`HELP` (`?`) gives a list of commands for which explanations are available to select from. Use `HELP` (the `HELP` heading of the menu bar) to obtain the dialog box shown, select the topic, and the `Go To` button **(Fig. 1.22)**.

The `SHELL` (`SH`), command gives you access to the DOS operating system while remaining in the `Drawing Editor` as follows:

`Command:` `SHELL` (ENTER)

`DOS command:` `DIR A:` `(or similar command)` (ENTER)

The `PURGE` command (under `Data`) can be used at any time to remove unused layers, blocks, and other attributes from files:

`Command:` `PURGE` (ENTER)

`Purge unused Blocks/Dimstyles/LAyers/LTypes/SHapes/`

`STyles/APpids/Mlinestyles/All:` `ALL` (ENTER)

`ALL` is used to eliminate all unused references one at a time as prompted. The other options purge specific features of a drawing.

The `Files Utilities` dialog box (under `File`, select `Management`, and `Utilities`) lists the following options: `LIST`, `COPY`, `RENAME`, `DELETE`, `UNLOCK`, and `HELP` **(Fig. 1.23)**.

`LIST` is used to obtain the `File List` box from which you can specify the drive (B:, for example) and obtain a list of the files in this drive **(Fig. 1.24)**.

FILES/MANAGEMENT/UTILITIES

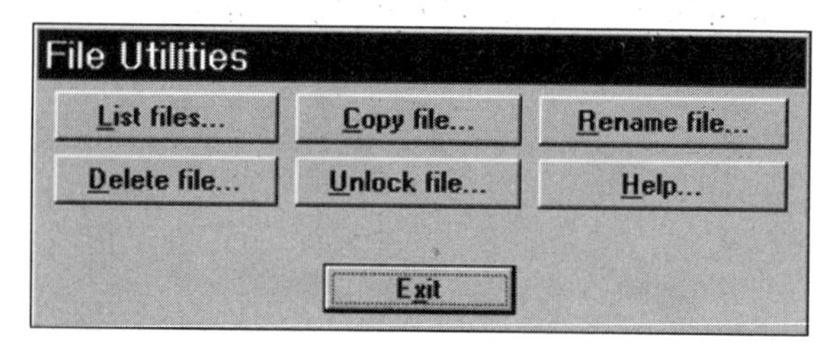

Figure 1.23 At the `Command` line, type `FILE` to get the `File Utilities` dialog box.

FILE/MANAGEMENT/FILE UTILITIES/FILE LIST

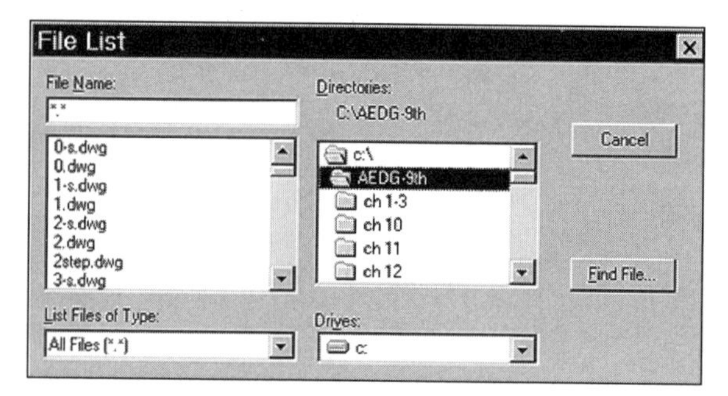

Figure 1.24 Select `List Files` from the `File Utilities` box (Fig. 1.23) to display the `File List` dialog box shown here. Files can be selected from any directory within this box.

`COPY` is picked to select a `Source file` and duplicate it to a `Destination file`.

`RENAME` lets you select `Old File Name` and rename it to a `New File Name`.

`DELETE` lets selected files be deleted (`Files-Delete-OK`). Specify a file, such as `B:DRAW1.DWG`; or use wild cards (`B:*.DWG`, for example), to delete all files with a `.DWG` extension. When using wild cards, files will be listed one at a time for deletion by entering `Y` or `N`.

`UNLOCK` is used to select `Files to Unlock` and pick `OK`.

LAYER CONTROL BOX (DDLMODES)

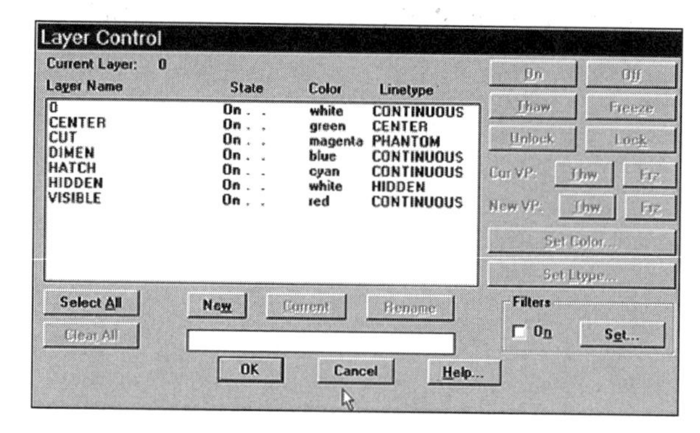

Figure 1.25 From the `Data` pull-down menu, select `Layers` to obtain this `Layer Control` dialog box for modifying existing layers and adding new ones.

1.7 Drawing Layers

An infinite number of layers can be created on which to draw, each assigned a name, color, and line type. For example, a yellow layer named `HIDDEN`, for drawing dashed lines, may be created.

Architects use separate copies of the same floor plan for different applications: dimensions, floor finishes, electrical details, and so forth. The same basic plan is used for all of these applications by turning on the needed layers and turning others off.

Setting Layers The layers shown in the `Layer Control` box (`DDLMODES`) in **Fig. 1.25** are sufficient for most working drawings. Layers are assigned line types and different colors so they can be easily distinguished from each other. The `0` (zero) layer is the default layer, which can be turned off or frozen but not deleted.

`LAYERS` can be created by typing a layer name in the box in Fig. 1.25 and selecting `New`. The new layer will appear in the listing of layers. Seven new layers have been assigned by name, each with a default color of white and a continuous line type, which can be changed.

SELECT COLOR BOX (DDCOLOR)

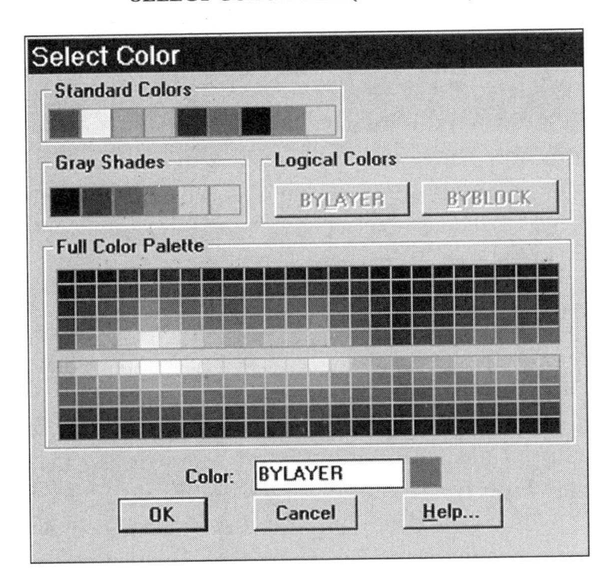

Figure 1.26 From the `Layer Control` dialog box (Fig. 1.25), select `Color` to obtain this `Select Color` dialog box for assigning colors to a layer.

LINETYPES

CONTINUOUS
BORDER
BORDER2
BORDERX2
CENTER
CENTER2
CENTERX2
DASHDOT
DASHDOT2
DASHDOTX2
DASHED
DASHED2
DASHEDX2
DIVIDE
DIVIDE2
DIVIDEX2
DOT
DOT2
DOTX2
HIDDEN
HIDDEN2
HIDDENX2
PHANTOM
PHANTOM2
PHANTOMX2

Figure 1.27 These linetypes are available for assignment to layers.

Color Pick a layer with the cursor and pick the `Set Color` button to assign it a color (Fig. 1.25). The `Select Color` dialog box appears on the screen from which a color can be chosen for the selected layer and the `OK` button picked **(Fig. 1.26)**.

Linetypes The `linetype`s and their names that are available are shown in **Fig. 1.27**. Lines on the `HIDDEN` layer will be drawn with dashed (hidden) lines. From the `Data` pull-down menu select `Linetype...` and the `Load...` button to get prompts for the lines to load **(Fig. 1.28)**. By picking `Layer` and the `Set Ltype` box in the `Layer Control` box (Fig. 1.25), these lines will appear for selection in a submenu.

Set A layer must be `SET` before it can become the current layer and be drawn. The current layer can be set by selecting it from the `Layer Control` dialog box and picking the `Current` button (Fig. 1.25).

LOAD OR RELOAD LINETYPES BOX

Load or Reload Linetypes
File... C:\r13\com\SUPPORT\acad.lin
Available linetypes
BORDER
BORDER2
BORDERX2
CENTER
CENTER2
CENTERX2
DASHDOT
DASHDOT2
DASHDOTX2
DASHED
DASHED2
DASHEDX2
Select All | Clear All
OK | Cancel | Help...

Figure 1.28 From the `Data` pull-down menu select `Linetype...` and the `Load...` to get these linetypes that can be loaded.

OBJECT PROPERTIES TOOLBAR

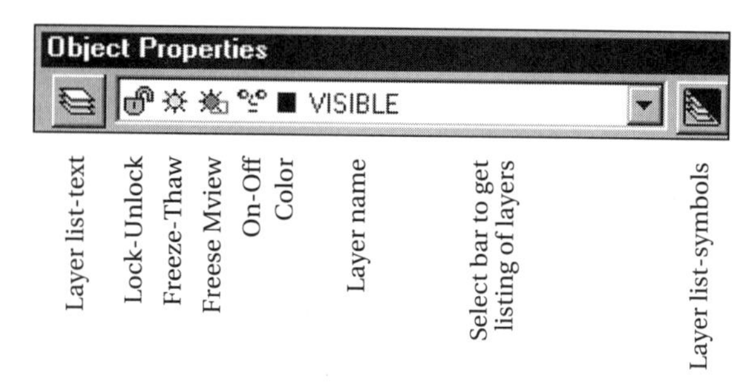

Figure 1.29 Layers can be seen and modified from the `Object Properties` toolbar at the top of the screen.

OBJECT PROPERTIES TOOLBAR

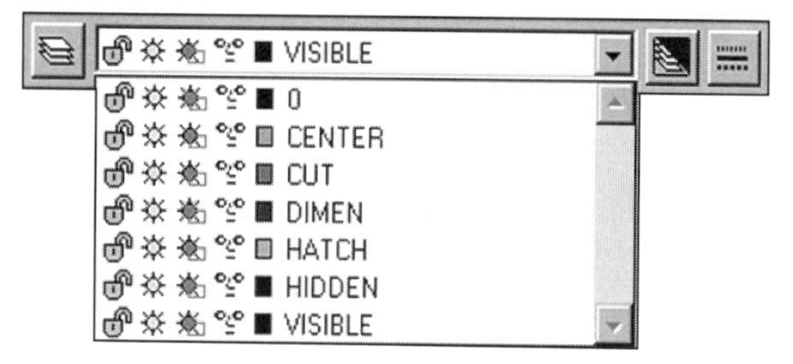

Figure 1.30 By selecting the bar with the layer symbols in the `Object Properties` toolbar, the drop-down list extends to show all the layers that have been set.

The `Object Properties` toolbar located at the top of the screen **(Fig. 1.29)** can be used to set the current layer and make other layer assignments. By selecting the `Layer List` (Text) icon at the left of the toolbar, the `Layer Control` dialog box is shown on the screen (Fig. 1.25). Select the `Layer List (Symbols)` icon at the right of the bar to get a list of symbols for all layers extended below the current layer **(Fig. 1.30)**.

`RENAME` changes a layer's name when the layer is selected from the dialog box, given a new name in the edit box, and the `RENAME` button is picked.

Drawings can be made only on the current layer, even if it is off. By typing `LAYER` (`LA`) and `(?)`, or by using the toolbar (Fig. 1.30), a listing of the layers, their linetypes, colors, and on/off status are displayed in either text or symbolic format.

`On/Off` of a layer is selected from the symbolic listing in the `Object Properties` toolbar (Fig. 1.30). When this icon is gray the layer is `Off`; when it is dark the layer is `On`. Layers can be turned `On` or `Off` with buttons from the `Layer Control` box (Fig. 1.25), or by typing `LAYER` and `Off` at the command line.

`Freeze` and `Thaw` options under the `LAYER` command are used like the `On` and `Off` options. `Freeze` a layer and it will be ignored by the computer until it has been `Thaw`ed, which makes regeneration faster than when `Off` is used. The current layer cannot be frozen whereas the current layer can be off.

Filters are used to sort layers by their properties specified in the `Layer Name` list (Fig. 1.25). Select `FILTERS` and the `Set Layer Filters` subdialog box appears for specifying the properties for sorting files. **Drop-down lists** show the options for `On/Off`, `Thaw/Freeze`, `Lock/Unlock`, or both. Layer `Names`, `Colors`, and `Ltypes` can be typed in the edit boxes of a certain color, name, or linetype that are to be listed in the `Layer Control` box when filtered (Fig. 1.25). Select `OK` to obtain the filtered list.

`LTSCALE` modifies the lengths of line segments and their spacings of noncontinuous lines, such as hidden lines.

Save Layers You should save these settings for future use with the `SAVE AS` command under the `File` menu. You can save `LAYERS`, `UNITS`, `GRID`, `SNAP`, `ORTHO`, and other drawing aids in a file (`FORMAT`, for example) to use as a `PROTOTYPE` (Fig. 1.4).

1.8 Toolbars

Toolbars are menus that appear in symbolic form as groups of icons that can be selected by cursor. A portion of the `Standard` toolbar **(Fig. 1.31)** gives a list of icons representing commands from `NEW` to `PASTE`. **Figure 1.32** shows additional commands from `UNDO` to `PRESET UCS`.

STANDARD TOOLBAR: PART 1

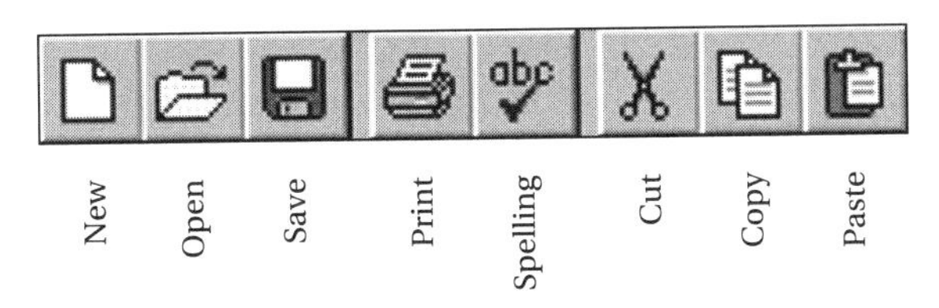

Figure 1.31 The icons and their definitions from the first part of the Standard toolbar, located at the top of the screen, are shown here.

EXAMPLES OF TOOLBARS

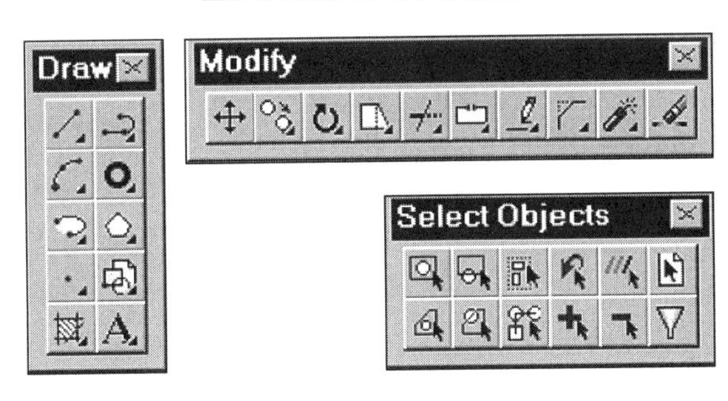

Figure 1.33 Examples of toolbars are shown here.

STANDARD TOOLBAR: PART 2

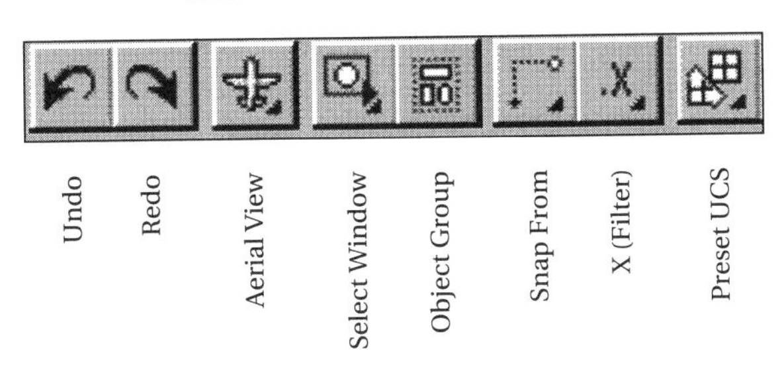

Figure 1.32 The icons and their definitions from the second part of the Standard toolbar, located at the top of the screen, are shown here.

OPENING A FILE

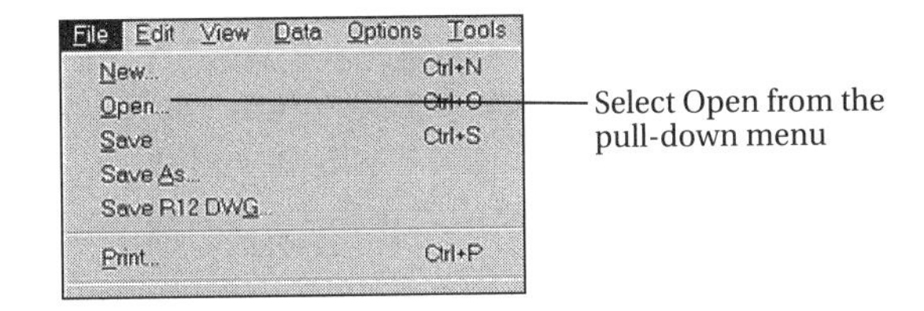

Figure 1.34 To open a file, select Open under File on the menu bar.

Toolbars are loaded by selecting the TOOLS menu of the menu bar, the Toolbar option, and the name of the toolbar (Fig. 1.14). Toolbars can be **moved** about the screen by selecting a point on one of their edges, holding the select button of the mouse down, and moving the cursor to a new position **(Fig. 1.33)**. Toolbars can be **docked** by moving them into contact with one of the perimeter borders of the screen. They will become vertical or horizontal strips depending on which border is selected. When positioned in the open area of the screen, toolbars will often appear as double-column menus. When a small icon is hard to read, place the pointer on an icon and a tooltip will appear with its definition.

1.9 Beginning a New Drawing (Standard Toolbar)

There are two ways of loading FORMAT containing its layers and settings. Begin with the Open or New options under File of the menu bar **(Fig. 1.34)**.

Method 1: Open it as an existing drawing, draw on it, and save it with SAVE AS to a different name (DWG1, for example) to create a new file with identical settings to FORMAT. SAVE AS makes DWG1 the current drawing on the screen and FORMAT remains unchanged on your disk. You should periodically back up DWG1 during a drawing session with the SAVE command from the pull-down menu, screen menu, toolbar, or keyboard.

CREATING A NEW DRAWING

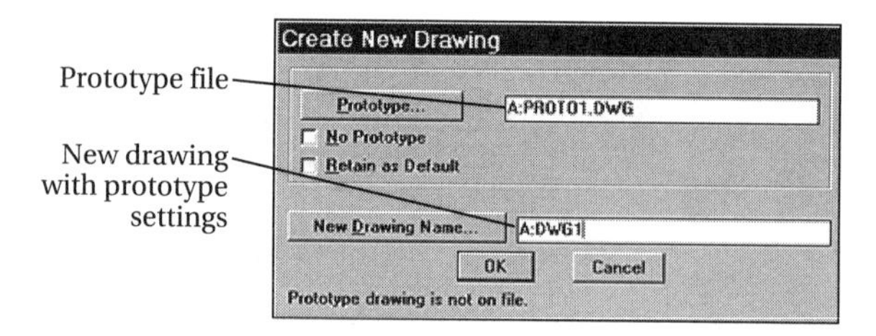

Figure 1.35 Under File, select New to obtain the Create New Drawing dialog box. Type the name of the prototype file, name the current drawing, and pick

SAVING A DRAWING

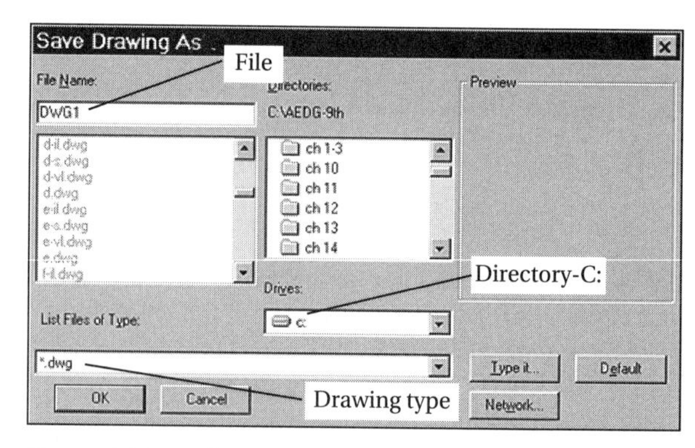

Figure 1.36 Under File, select SAVE AS to obtain the Save Drawing As dialog box where the directory can be selected and the file named NEW-2.

Method 2: Select New, type FORMAT as the PROTOTYPE file, type the new drawing's name (DWG1) in the edit box, and select OK **(Fig. 1.35)**. This procedure assigns the settings of FORMAT as the PROTOTYPE to a drawing named DWG1 without having to reset Layers and Drawing Aids.

DISCARDING CHANGES

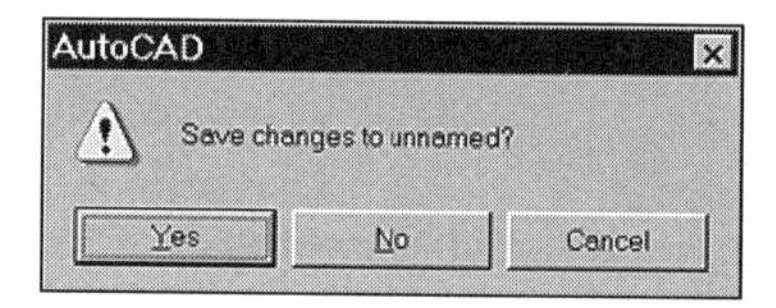

Select NO to exit without saving changes

Figure 1.37 If the current file has not been saved prior to selecting EXIT, this box will appear on the scene to allow you to save, not save, or cancel the command.

1.10 Saving and Exiting

The File pull-down menu gives options for saving a drawing—SAVE, SAVE AS, and EXIT—which can be selected from the pull-down menu, the screen menu, the toolbar, or by typing at the command line.

From the Menu Bar

Use the SAVE command to "quick save" to the current file's name if the file has been previously named and saved. If the file is unnamed, the SAVE command prompts for a file name by displaying the Save Drawing As dialog box in **Fig. 1.36.** Select the directory and name the file (NEW-2) to save it on the disk.

The SAVE AS command displays the Save Drawing As dialog box for verifying the current file name or to save the drawing under a different name. When a file is saved to a new name (NEW-2, for example), it becomes the current file on the screen.

The EXIT command gives the dialog box shown in **Fig. 1.37,** which prompts for the file name unless you have just saved it. To exit without saving, respond to Save Changes to NEW-2 with NO to discard the changes made since the last SAVE.

Ending the Session

Select EXIT from the pull-down menu or type END or QUIT to exit from AutoCAD if your work has been saved.

END saves the current drawing and exits AutoCAD whether or not any changes have been

PLOT CONFIGURATION SPECIFICATIONS

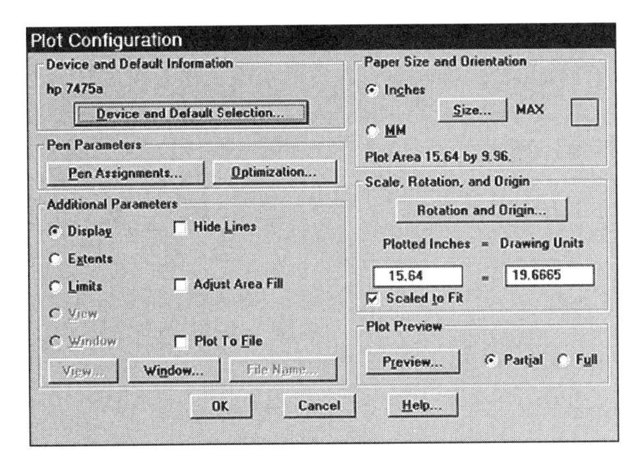

Figure 1.38 Under `File`, pick `PRINT` to get this `Plot Configuration` dialog box for plotting a file.

DEVICE AND DEFAULT SELECTION BOX

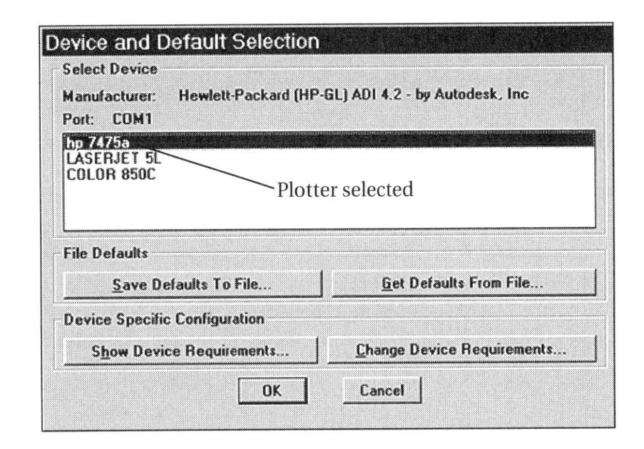

Figure 1.39 By selecting `Device and Default Selection` (Fig. 1.38) this dialog box appears, from which the plotter/printer device can be specified.

PCP FILE (PLOT CONFIGURATION)

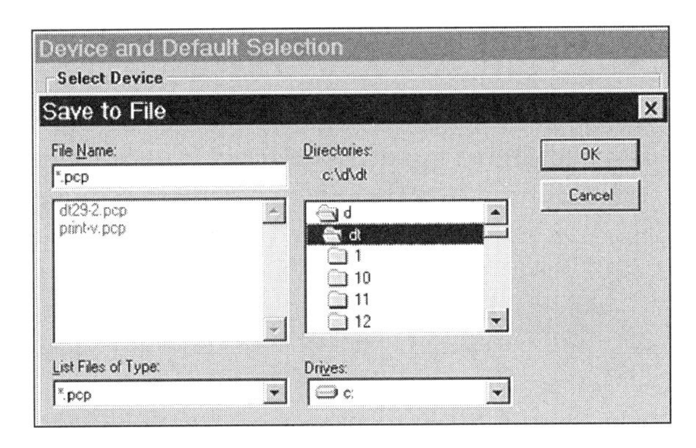

Figure 1.40 A `Plot Configuration Parameters` (`.PCP`) file of various settings can be saved for repetitive usage by selecting `Save Defaults to File` (Fig. 1.39) and saving as a `.PCP` file.

made. If the drawing has not been previously named, the `Create Drawing File` dialog box will prompt you for a file name. The previous version of the drawing is automatically saved as a backup file with a `.BAK` extension and the current drawing is saved with a `.DWG` extension.

`QUIT` exits AutoCAD if changes have been saved, or it gives a dialog box shown in Fig. 1.37 with three options if there are unsaved changes. If `Yes` is selected and the file is named, the program exits from AutoCAD. The `EXIT` command under the `File` pull-down menu has the same function as the `QUIT` command.

1.11 Plotting Parameters

You can plot a drawing before ending a drawing session by selecting `PRINT` under the `File` heading of the menu bar to obtain the `Plot Configuration` dialog box **(Fig. 1.38)**. Select the `Device and Default Selection` button to obtain the subdialog box **(Fig. 1.39)** for selecting `PLOTTER` to make a pen drawing.

Select the `Save Defaults to File` button to display the `Save to File` subdialog box for creating a file with a `.PCP` extension called a `Plot Configu-ration Parameters` file **(Fig. 1.40)**. Name this file `DWG1` (it is given a .PCP extension automatically) since it will be used in conjunction with the file `DWG1.DWG`.

`Show Device Requirements` is selected in Fig. 1.39 to obtain the dialog box for specifying how

PEN ASSIGNMENTS DIALOG BOX

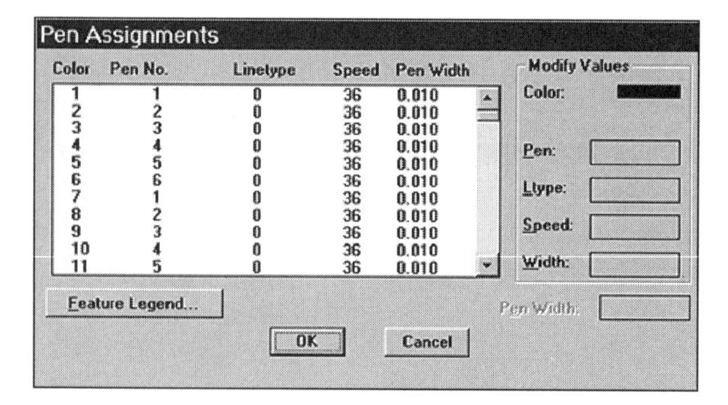

Figure 1.41 Select the `Pen Assignments` option from the `Plot Configuration` dialog box, which shows color, pen number, linetypes, speed, and pen width.

ADDITIONAL PARAMETERS

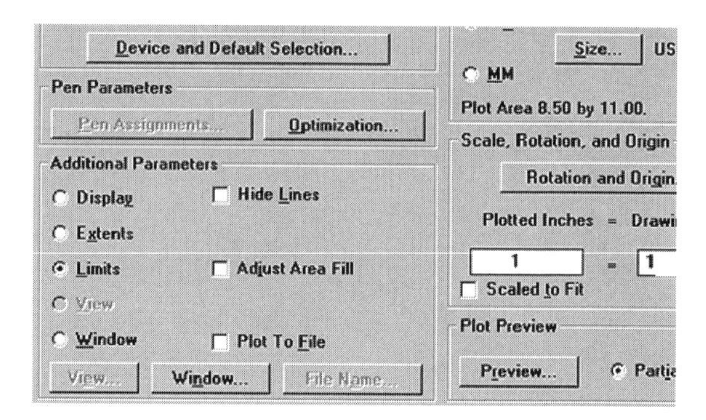

Figure 1.42 The `Additional Parameters` area is used to specify the area to be plotted and to hide lines in 3D drawings.

long the plotter will wait for plotter port time-outs and other requirements. Pick `OK` to leave this box. The `Change Device Requirements` box (Fig. 1.39) is chosen to assign the plotting device; then pick `OK`.

The `Pen Assignments` box shows the assignment of colors to pen numbers (slots in pen holder on the plotter) **(Fig. 1.41)**. For example, a red line (`VISIBLE` layer) will be plotted with the pen in slot 1, a P.7 pen.

`Ltype` is set to `0` since the layers have had `linetypes` assigned to them. `Pen` assigns pens to the colors on the screen. `Speed` designates the rate at which the plotter pen moves. A speed of 9 inches per second is average and 36 is the maximum. `Width` gives the line widths based on pen tip widths for drawing multistroke lines (polylines or traces). Use the default width of 0.01 for plotters. Line widths are assigned when laser printers are used and this option is grayed out if not supported by a printer.

`Feature Legend`, a subdialog box of the `Pen Assignments` box, shows linetypes for selection. Do not use this option when layers have been assigned linetypes as part of the drawing file.

`Optimization` of the `Plot Configuration` dialog box (Fig. 1.38) shows a dialog box for reducing pen motion. When a box is selected, those above it become active except for the `No optimization` box. The last two boxes pertain to 3D figures to eliminate multiple strokes where lines coincide.

The `Additional Parameters` area of the `Plot Configuration` dialog box **(Fig. 1.42)** offers the following settings:

`Display` plots the drawing as shown on the screen.

`Extents` plots a drawing to its extents if the scale selected permits. It is good practice to apply `Zoom/Extents` to ready a drawing for plotting.

`Limits` is used to plot the portion of the drawing lying within the grid pattern as defined by typing `LIMITS`.

`View` lists saved `Views` that can be selected and plotted. This box is gray if no views are saved.

`Window...` displays a `Window Selection` subdialog box which lets you define a rectangular portion of a drawing to be plotted. Press the `Pick <` button to define this window with the cursor or type coordinates in the edit boxes **(Fig. 1.43)**.

`Hide Lines` removes hidden lines from 3D drawings that are plotted.

WINDOW SELECTION

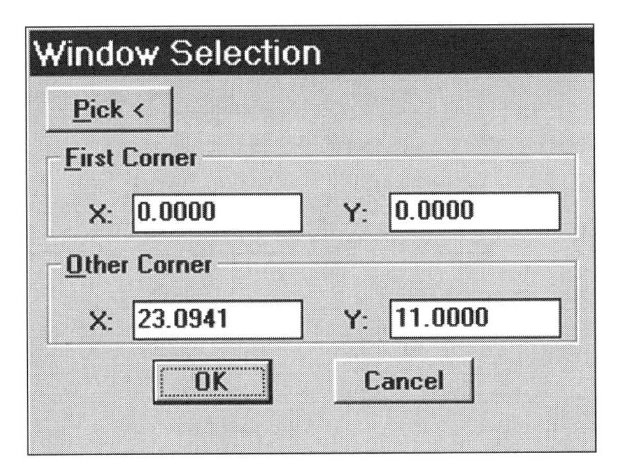

Figure 1.43 When `Window` is selected from the `Additional Parameters` area of the `Plot Configuration` dialog box, the coordinates of the diagonal corners of the window to be plotted can be `PICK`ed or specified.

`Adjust Area Fill` pulls in the boundaries of the filled area by one-half pen width for a more precisely drawn fill area.

`Plot to File` sends the plot to a file with a `.PLT` extension for plotting with a utility program instead of the plotter.

`File Name...` gives a dialog box for selecting a plot file.

The `Paper Size and Orientation` area is used to specify inches or millimeters (MM) as units for a plot **(Fig. 1.44)**. A rectangle in this area denotes whether the page has a portrait (vertical) or landscape (horizontal) orientation.

`Size...` displays a `Paper Size` sub-dialog box listing standard and user-specified plot sizes **(Fig. 1.45)**. `MAX` is the largest size that the selected plotter can handle. When a size is selected and `OK` is picked, the sub-dialog box disappears, and the size is shown in the `Plot Area` (Fig. 1.44).

PAPER SIZE AND ORIENTATION

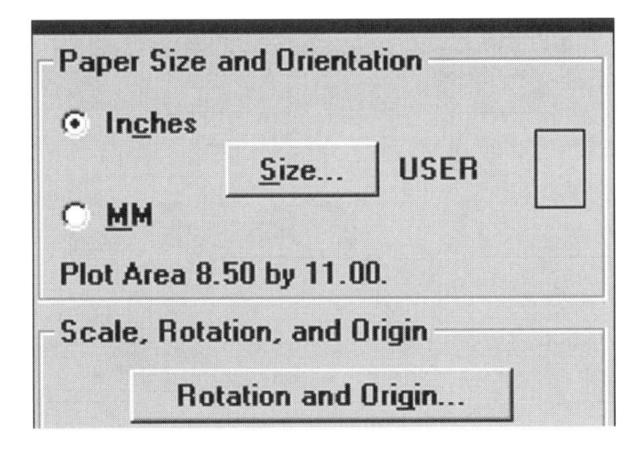

Figure 1.44 From the `Paper Size and Orientation` area, select `inches` or `MM` as the units and select `Size...` (size A here) to obtain a listing of plot sizes.

PAPER SIZE AND ORIENTATION

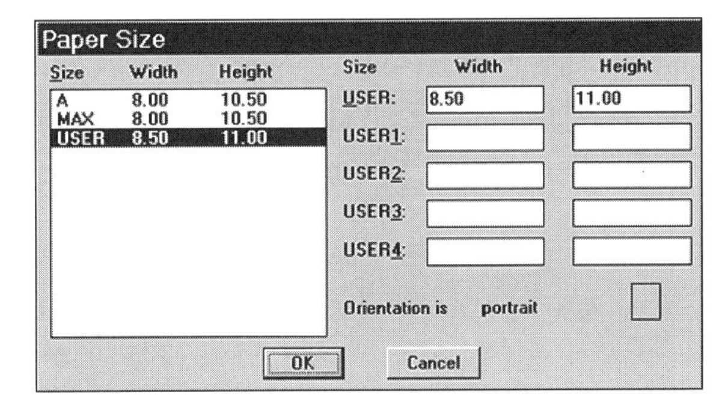

Figure 1.45 The `Paper Size` dialog box gives a listing of the plot sizes that are available. User sizes for plots can be specified in the boxes at the right.

The `Rotation and Origin...` button in the `Scale, Rotation, and Origin` area displays a `Plot Rotation and Origin` sub-dialog box **(Fig. 1.46)**. `Plot Rotation` radio buttons rotate the plots 0, 90, 180, or 270 degrees. `Plot Origin` is usually specified as 0,0 (the lower left corner of the sheet), but it can be offset by X- and Y-values.

PLOT ROTATION AND ORIGIN DIALOG BOX

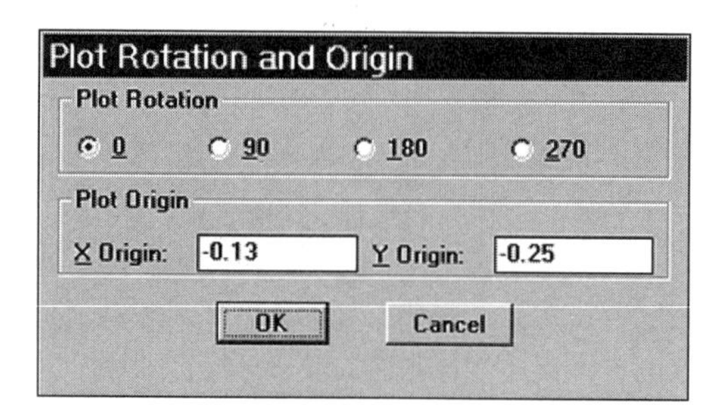

Figure 1.46 Select Rotation and Origin to display this box for specifying the rotation of the plot and its X- and Y-offsets.

PARTIAL PREVIEW OF A PLOT

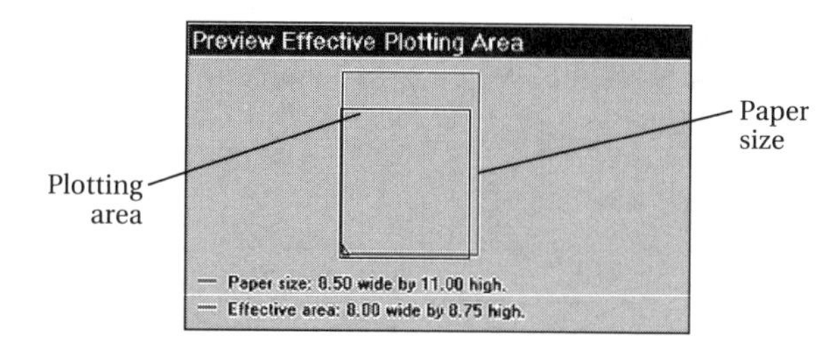

Figure 1.48 Select Partial Preview to obtain two rectangles on the screen that represent the paper size and the plotting area.

SCALING TO FIT SHEET SIZE

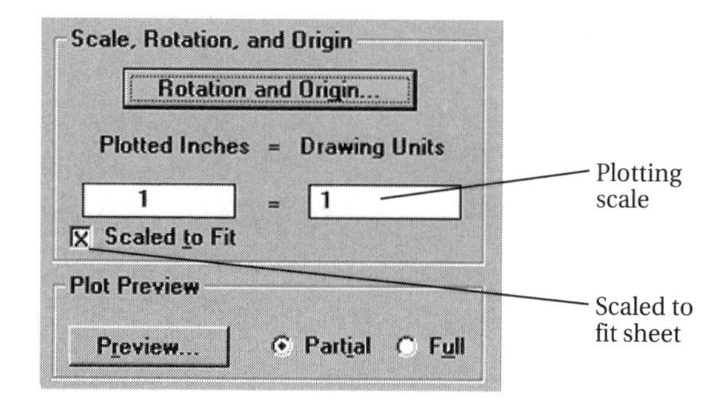

Figure 1.47 Insert the scale of a plot in the boxes of the Plot Configuration box. 1 = 2 is half size, and 2 = 1 is double size.

Scale is typed in the edit boxes, Plotted Inches = Drawing Units **(Fig. 1.47)**. If metric units had been selected, the label would read Plotted MM. 1 = 1 is full size, 1 = 2 is half size, and 2 = 1 is double size.

Scaled to Fit makes the drawing's extents fill the plotting area and be as large as possible.

Plot Preview gives partial and full previews. Partial Preview shows rectangles representing paper size and the drawing area **(Fig. 1.48)**. The part of the drawing area that exceeds the paper size cannot be plotted unless the scale, origin, or both are adjusted. The triangular rotation icon is shown in the lower left corner for zero rotation, upper left for 90 degrees of rotation, upper right for 180 degrees of rotation, and lower right for 270 degrees of rotation.

Full Preview shows the entire drawing on the screen and its relationship to the paper limits when plotted **(Fig. 1.49)**. Pan and Zoom can be performed to check a drawing for details during preview.

End Preview returns to the Plot Configuration dialog box and OK to plot the drawing.

To Update the .PCP file, pick Save Defaults to File... as shown in **Fig. 1.50** to save these settings and make them available from the Get Defaults From File for a plot with the same settings.

1.12 Readying the Plotter

After the parameters are set and the preview is approved, pick OK and the Command line gives the following messages at the bottom of the screen:

FULL PREVIEW OF A PLOT

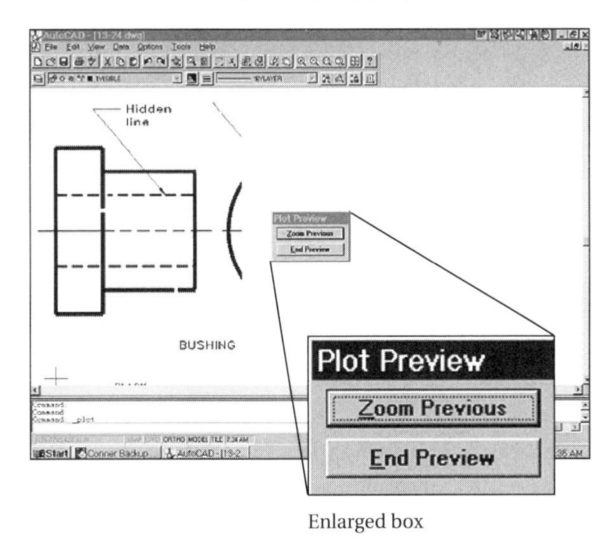

Figure 1.49 Select `Full Preview` to obtain the paper size and a full view of the drawing to be plotted. The sub-dialog box can be used to `Pan` and `Zoom` the drawing. Select `End Preview` and `OK` to plot the drawing.

PLOT CONFIGURATION PARAMETERS (.PCP)

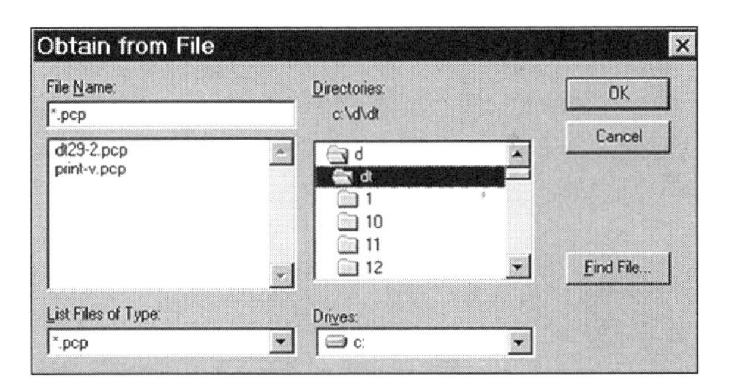

Figure 1.50 To save the previous settings for future use, save the file as a `.PCP` file by using this dialog found under the `Device and Default Selection` box of the `Plot Configuration` dialog box.

`Effective plotting area: XX wide by YY high` (ENTER)

`Position paper in plotter.` (ENTER)

7475A HP PLOTTER

1. Load paper, lower lever
2. Pens: P.7 (Slot 1); P.3 (Slot 2)
3. Set for A-Size
4. Press P1 and P2

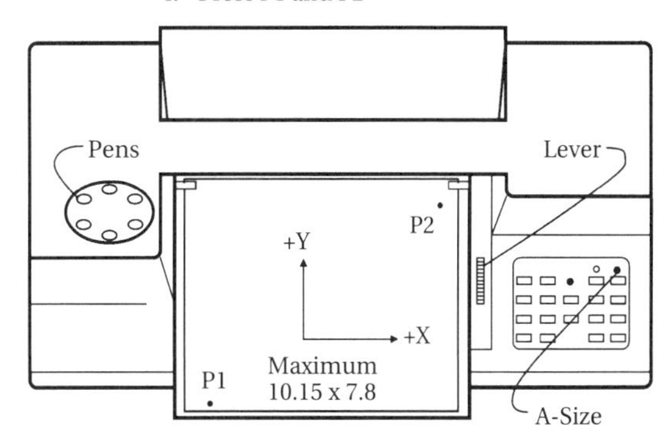

Figure 1.51 This Hewlett-Packard 7475A plotter is typical of those used to plot A and B sizes.

`Press` (ENTER) `to continue or S to Stop for hardware setup`

Load paper in the plotter as shown in **Fig. 1.51** with the thick pen (P.7) in slot 1 and the thin pen (P.3) in slot 2 as specified by `Pen Assignments` in Fig. 1.41. Press (ENTER) and the plot will begin. Plotting can be cancelled by pressing `Esc`, but it may take almost a minute for it to take effect. When completed, select `EXIT` from `File` on the menu bar to end the session and leave AutoCAD.

Now that we know how to set a few drawing aids, save files, and plot, it is time to learn how to make drawings.

1.13 2D LINES (Draw Toolbar)

`Open` your prototype drawing, `FORMAT`, and use `SAVE AS` to name the drawing `N01`, which becomes the current drawing with the settings of `FORMAT`. Select `TOOLS`, the `Toolbars option`, and `Draw` to load the `Draw` toolbar onto the screen **(Fig. 1.52)**.

A `Line` (called an object) can be drawn by using the screen menu, the keyboard, or the `Draw`

THE DRAWING TOOLBAR

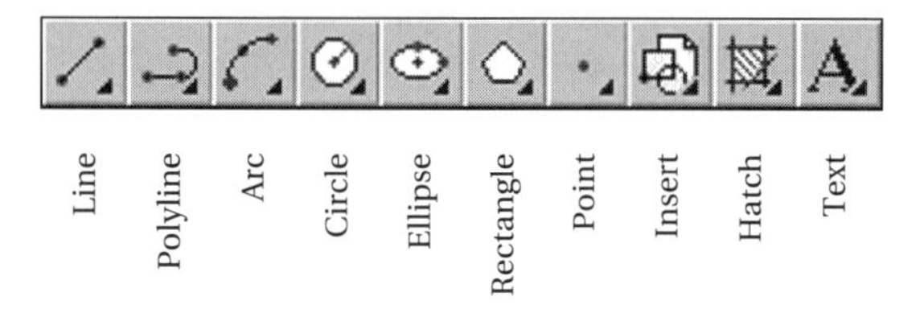

Figure 1.52 From under TOOLS on the menu bar, select Toolbars, and pick Draw to load this toolbar for selecting the commands above.

LINE (DRAW TOOLBAR)

Line Construction line Ray

Figure 1.54 Icons in the toolbar with small black triangles have flyouts with alternate commands as shown here. Hold the select button down on the icon to obtain the flyouts.

DRAWING LINES

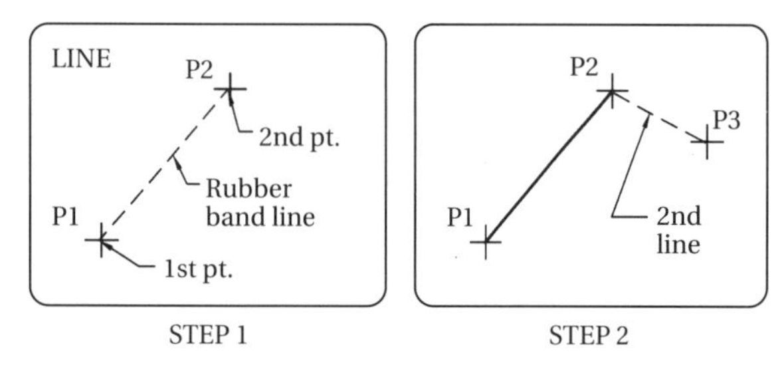

Figure 1.53 Drawing a LINE.
Step 1 Command: LINE (ENTER)
LINE From point: P1
To point: P2 (The line is drawn.)
Step 2 To point: P3 (The line is drawn.) ((ENTER) to disengage the rubber band.)

ABSOLUTE VS. POLAR COORDINATES

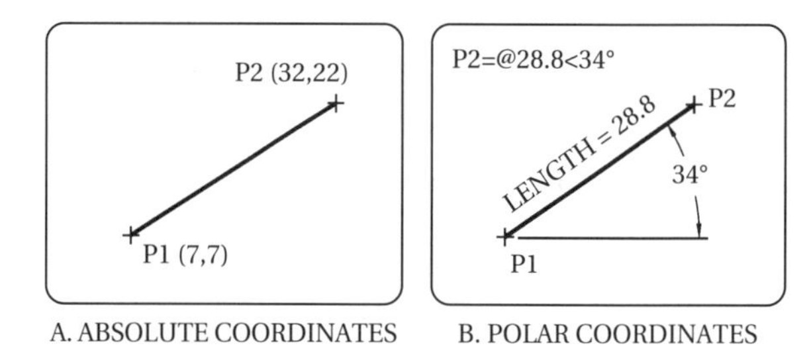

Figure 1.55
A Absolute coordinates can be typed to establish ends of a line.
B Polar coordinates are relative to the current point and are specified with a length and the angle measured counter-clockwise from the horizontal.

toolbar. Select DRAW1 and LINE (or L) from the screen menu, and respond to the command line as shown in **Fig. 1.53**, to draw lines by picking endpoints with the left button of the mouse. The current line will rubber band from the last point and lines are drawn in succession until (ENTER) or the right button of your mouse is pressed.

The Draw toolbar can be used to select the Line icon to get three flyout icons for types of lines: Lines, Construction Lines, and Rays **(Fig. 1.54)**. A **construction line** is drawn totally across the screen and a **ray** is drawn from the selected point to the edge of the screen.

A comparison of **absolute** and **polar** coordinates is shown in **Fig. 1.55**. **Delta** coordinates can be typed as @2,4 to specify the end of a line 2 units in the X-direction and 4 units in the Y-direction from the current end.

Polar coordinates are 2D coordinates and are typed as @3.6<56 to draw a 3.6 long line from the current (and active) end of a line at an angle of 56° with the X-axis.

Last coordinates are found by typing @ while in the LINE command. This causes the cursor to move to the last point.

World coordinates locate points in the **World**

POINT (DRAW TOOLBAR)

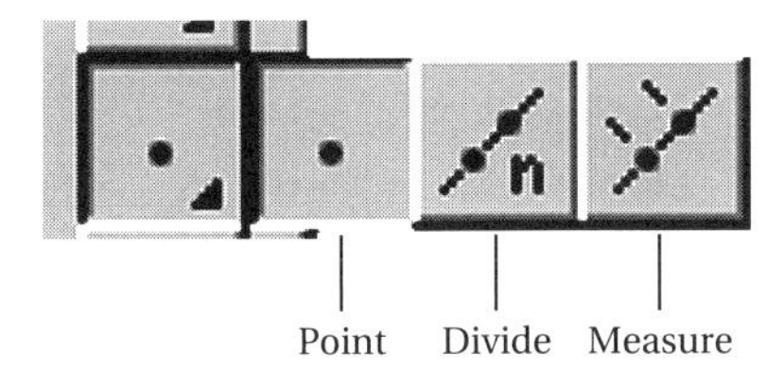

Figure 1.56 `Point` is one of the flyouts you can access from the `Draw` toolbar.

POINT STYLE BOX (DDPTYPE)

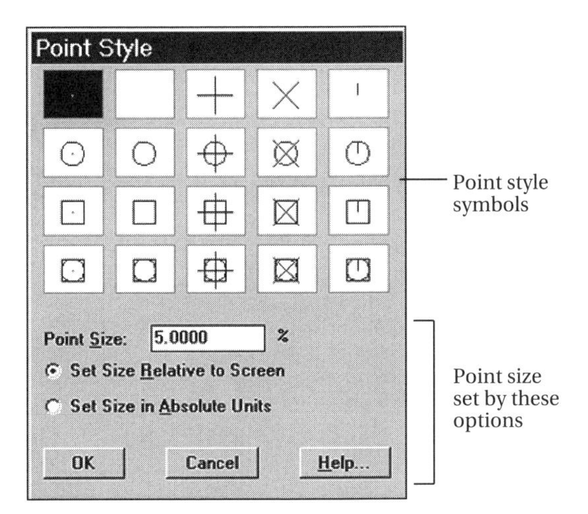

Figure 1.57 From `Options`, select `Display` and then `Point Style` to obtain this box for assigning point styles and sizes of point markers. You can also type `DDPTYPE` at the `Command` line to get this box.

Coordinate System regardless of the **User Coordinate System** being used by preceding the coordinates with an asterisk (*). Examples are `*4,3`; `*90<44`; and `@*1,3`.

Errors can be corrected while typing commands by the following methods:

CTRL X — Deletes the line.

ESC — Cancels the current command and returns the `Command:` prompt.

CIRCLE (DRAW TOOLBAR)

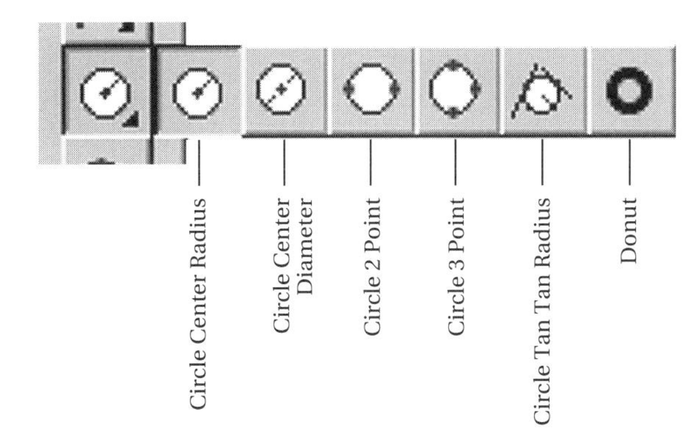

Figure 1.58 From the `Draw` toolbar, flyouts are available for drawing six types of circular features.

BACKSPACE — Deletes one character at a time.

The status line at the bottom of the screen shows the length of the line and its angle from the last point as it is being rubber-banded from point to point. The `CLOSE` command will close a continuous series of lines from the last to first point selected.

1.14 Points (Draw Toolbar)

The `POINT` command can be selected from the `Draw` toolbar **(Fig. 1.56)**. To obtain the menu of point symbols in **Fig. 1.57**, type `DDPTYPE`. Select a point `Style` and `Point Size` to set its size either **relative** to the screen or in **absolute units**. Unlike `Blips`, `Points` will plot as part of the drawing.

1.15 Circles (Draw Toolbar)

The `CIRCLE` command under `DRAW1` of the screen menu will draw circles when you select a center and radius, a center and diameter, or three points. Typing `C` at the `Command` line is the fastest means of activating the `CIRCLE` command, but the circle icon on the `Draw` toolbar **(Fig. 1.58)** can also be selected. **Figure 1.59** illustrates how a circle is drawn.

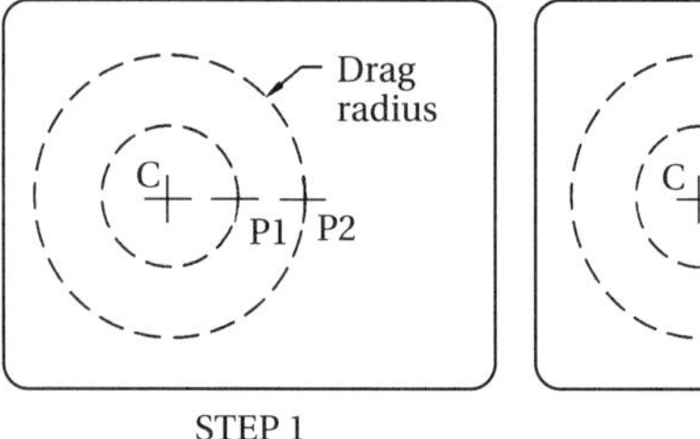

Figure 1.59 Drawing a `CIRCLE`.
Step 1 `Command:` <u>`CIRCLE`</u> (ENTER)
`2P/2P/TTR/<Center point>:` <u>`C`</u> (with cursor)
`Diameter/<Radius>:` (Drag radius to P1, and then to P2 to dynamically change it.)
Step 2 The final radius is selected at `P2` to draw the circle.

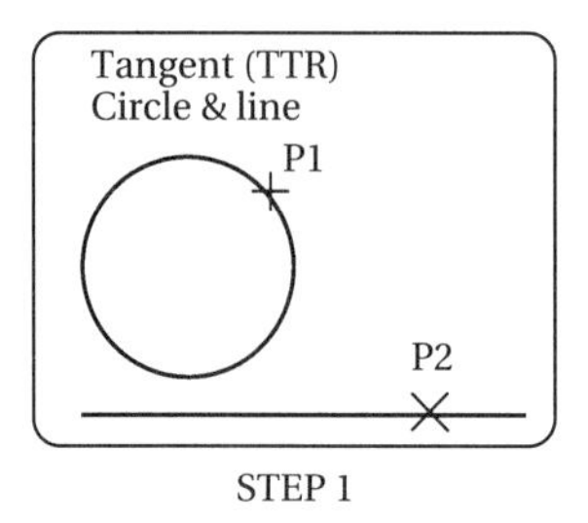

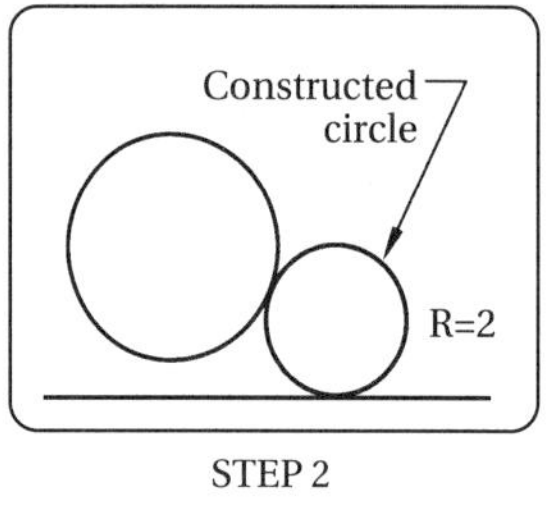

Figure 1.60 Circle: Tangent options (`TTR`).
Step 1 `Command:` <u>`CIRCLE`</u> (ENTER)
`3P/2P/TTR/<Center Pt>:` <u>`TTR`</u> (ENTER)
`Enter Tangent spec:` <u>`P1`</u>
`Enter second Tangent spec:` <u>`P2`</u>
Step 2 `Radius:` <u>`2`</u> (ENTER)
(The circle with radius = 2 is drawn tangent to the line and circle.)

The `TTR` (tangent, tangent, radius) option of the `CIRCLE Command` draws a circle tangent to a circle and a line, three lines, three circles, or two lines and a circle. A circle is drawn tangent to a line and a circle by selecting the circle, the line, and giving the radius **(Fig. 1.60)**.

The `3P` option calculates the radius length and draws a circle tangent to three lines **(Fig. 1.61)**.

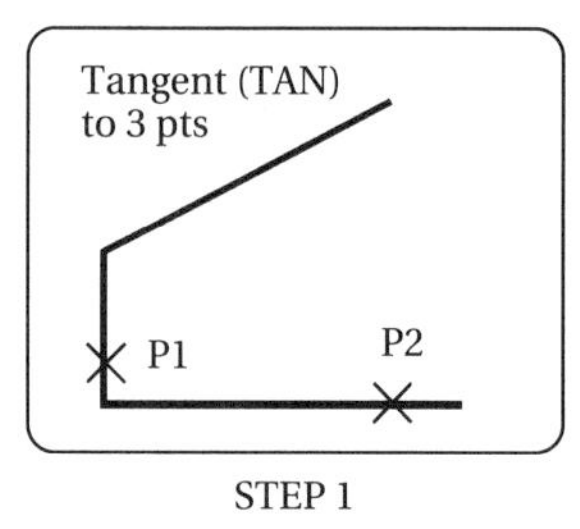

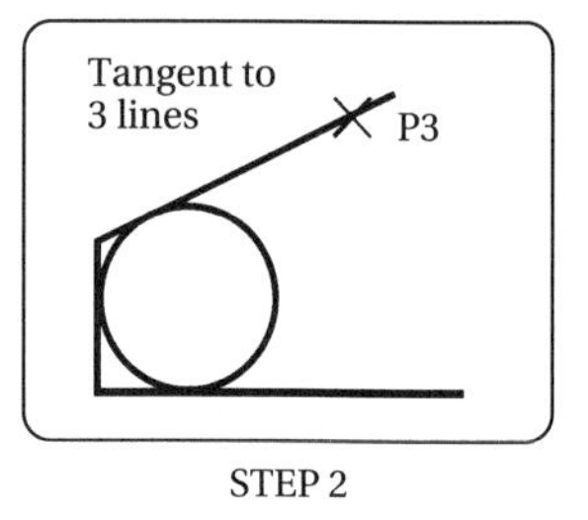

Figure 1.61 Circle: Tangent to three lines.
Step 1 `Command:` <u>`CIRCLE`</u> (ENTER)
`3P/2P/TTR/ <Center point>:` <u>`3P`</u>
`First point:` <u>`TAN`</u> (ENTER) `to` <u>`P1`</u>
`Second point:` <u>`TAN`</u> (ENTER) `to` <u>`P2`</u>
Step 2 `Third point:` <u>`TAN`</u> (ENTER) `to` <u>`P3`</u>
(The circle is drawn tangent to three lines.)

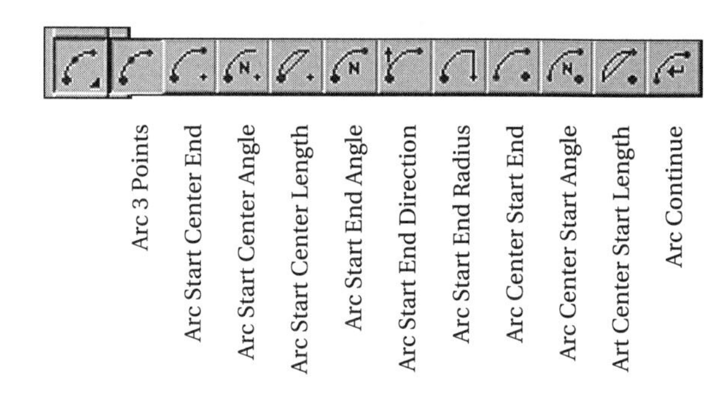

Figure 1.62 From the `Draw` toolbar, flyouts are available for drawing eleven types of arcs.

1.16 Arcs (Draw Toolbar)

The `ARC` command, under `DRAW1` of the screen menu or in the `Draw` toolbar **(Fig. 1.62)**, has eleven combinations of variables that use abbreviations for starting point, center, angle, ending point, length of chord, and radius. The `S`, `C`, `E` version requires that you locate the starting point `S`, the center `C`, and the ending point `E` **(Fig. 1.63)**.

DRAWING AN ARC

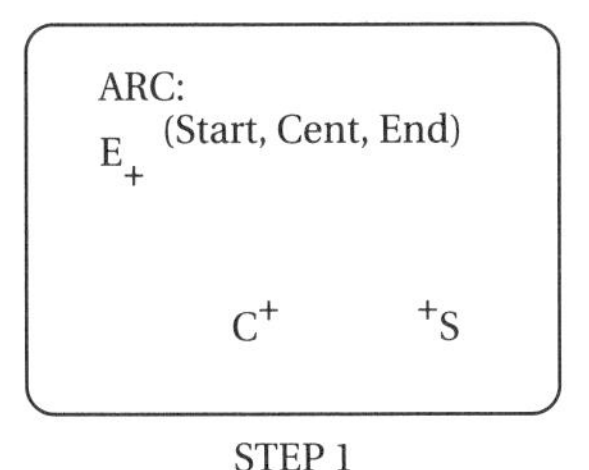

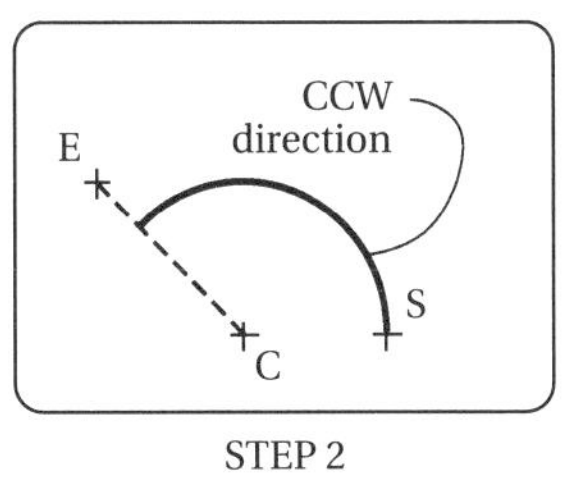

Figure 1.63 `ARC`: SCE option.
Step 1 `Command:` <u>`ARC`</u> (ENTER)
`Arc Center/<Start point>:` <u>`S`</u> (ENTER)
`Center/End/<Second point>:` <u>`CENTER or C`</u> (ENTER)
Step 2 `Angle/Length of chord/<Endpoint>:` <u>`DRAG to E`</u> (The arc is drawn CCW to an imaginary line from `S` to `E`.)

DRAWING RUNOUTS

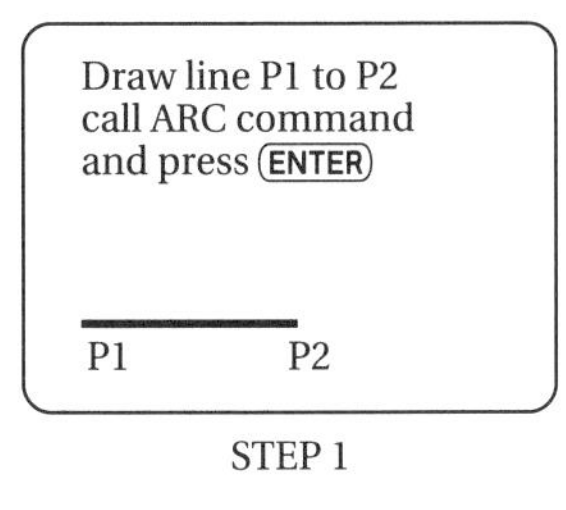

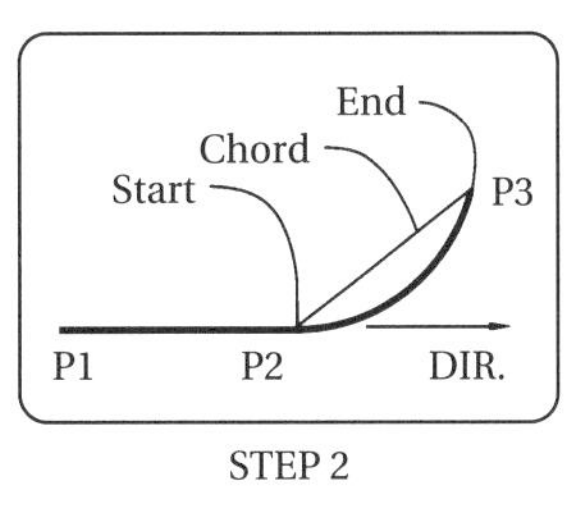

Figure 1.64 Arc tangent to end of line.
Step 1 `Command:` <u>`LINE`</u> (ENTER)
`From point:` <u>`P1`</u>
`To point:` <u>`P2`</u> (ENTER)
Step 2 `Command:` <u>`ARC`</u> (ENTER)
`Center/<Start point>:`(ENTER)
`Endpoint:` <u>`P3`</u>

The arc begins at point `S` and is drawn counterclockwise by default to a point near `E`.

A line can be continued with an arc drawn from the last point of a line and tangent to it as shown in **Fig. 1.64** for drawing runouts of fillets and rounds. It can be used to draw a tangent line from an arc by applying the commands in reverse and dragging the line to its final size.

FILLET COMMAND (MODIFY TOOLBAR)

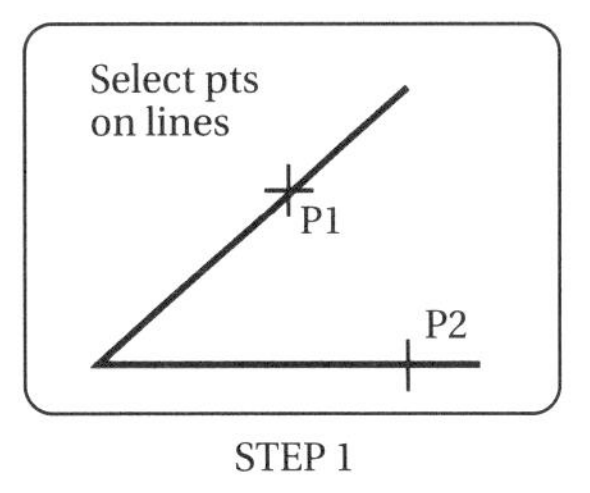

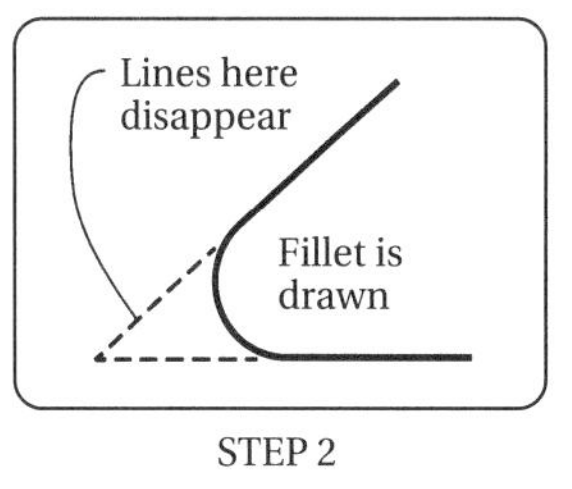

Figure 1.65 `FILLET` command.
Step 1 `Command:` <u>`FILLET`</u> (ENTER)
`Polyline/ Radius/Trim/<Select two objects>:` <u>`R`</u> (ENTER)
`Enter fillet radius <0.0000>:` <u>`1.5`</u> (ENTER)
`Command:` (ENTER)
Step 2 `FILLET Polyline Radius/<Select two objects>:` <u>`P1`</u> `and` <u>`P2`</u> (The fillet is drawn and the lines trimmed.)

1.17 Fillet (Modify Toolbar)

The corners of two lines can be rounded with the `FILLET` command whether or not they intersect. When the fillet is drawn, the lines are either trimmed or extended as shown in **Fig. 1.65**.

The assigned radius is remembered until it is changed. By setting the radius to 0, lines will be extended to a perfect intersection. Fillets of a specified radius can be drawn tangent to circles or arcs as shown in **Figs. 1.66** and **1.67**.

1.18 Chamfer (Modify Toolbar)

The `CHAMFER` command draws angular bevels at intersections of lines or polylines. After assigning chamfer distances (`D`), select two lines and they are trimmed or extended, and the `CHAMFER` is drawn **(Fig. 1.68)**. Press (ENTER) to repeat this command using the previous settings.

1.19 Polygon (Draw Toolbar)

An equal-sided polygon can be drawn as shown in **Fig. 1.69** with the `POLYGON` command by selecting the `Polygon` icon from the `Draw` toolbar **(Fig. 1.70)**.

FILLET: TANGENT ARCS

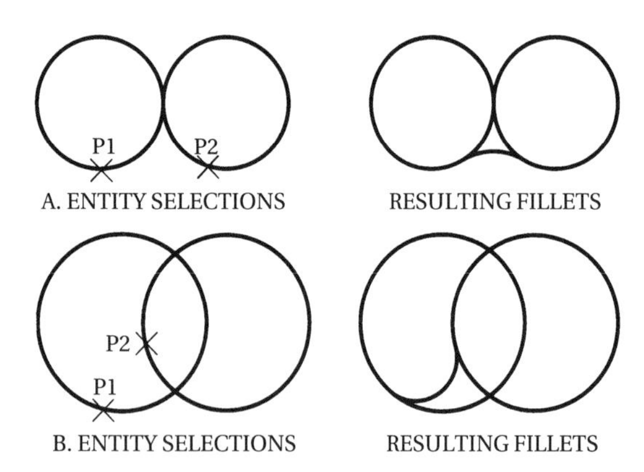

Figure 1.66 Tangent arcs.
Step 1 `Command:` <u>FILLET</u> (ENTER)
`Polyline/Radius/Trim/<Select two objects>:` <u>R</u> (ENTER)
`Enter fillet radius <0.0000>:` <u>1.2</u> (Example.) (ENTER)
Step 2 `Command:` (ENTER)
`FILLET Polyline/Radius/<Select two objects>:` (Select a point on each circle, and the fillet is drawn.)

FILLETS: LINES & ARCS

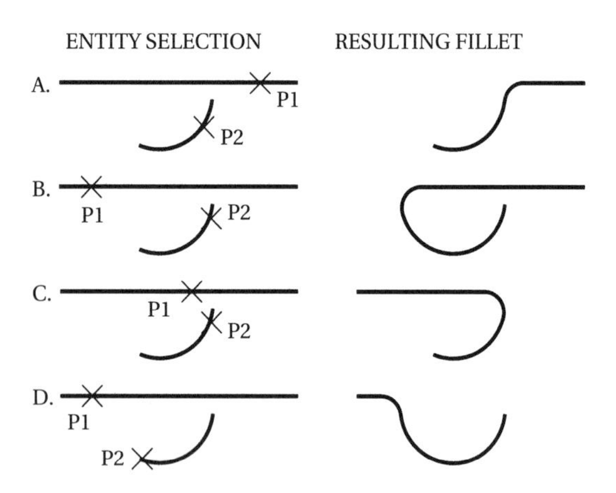

Figure 1.67 When a fillet is applied to a line and an arc, each fillet is determined by the location of the points selected.

CHAMFER COMMAND (MODIFY TOOLBAR)

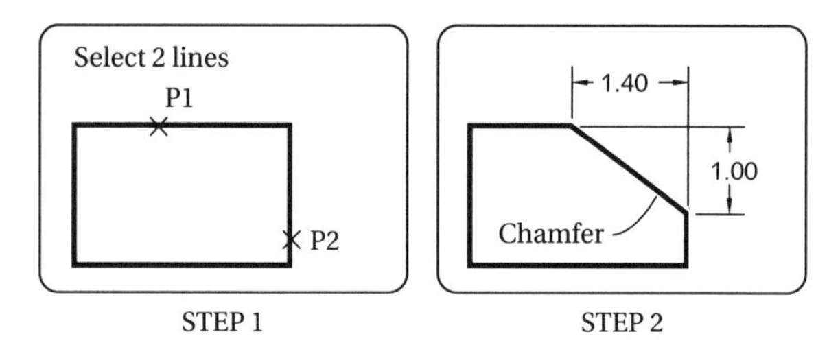

Figure 1.68 CHAMFERing a corner.
Step 1 `Command:` <u>CHAMFER</u> (ENTER)
`Polyline/Distance/Angle/Trim/Method/<Select first line>:` <u>D</u> (ENTER)
`Enter first chamfer distance <0.0000>:` <u>1.40</u> (ENTER)
`Enter second chamfer distance (0.000>:` <u>1.00</u> (ENTER)
`Command:` (ENTER)
`Polyline/Dist.../<Select first line>:` <u>P1</u> (ENTER)

POLYGONS

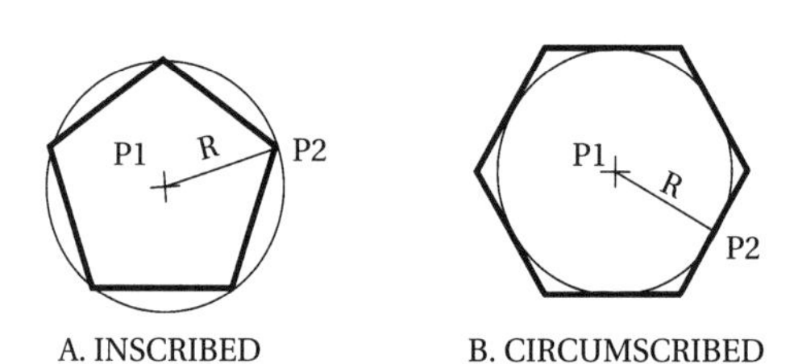

Figure 1.69 The POLYGON command can be used to draw polygons **inscribed** inside or **circumscribed** outside a circle.

`Command:` <u>POLYGON</u> (ENTER)
`Number of sides:` <u>5</u> (ENTER)
`Edge/<Center of polygon>:` (Locate center.)
`Inscribed in circle\Circumscribed about circle (I/C):` <u>I</u> (ENTER)
`Radius of circle:` (Type length and (ENTER) or select length with cursor.)

POLYGON (DRAW TOOLBAR)

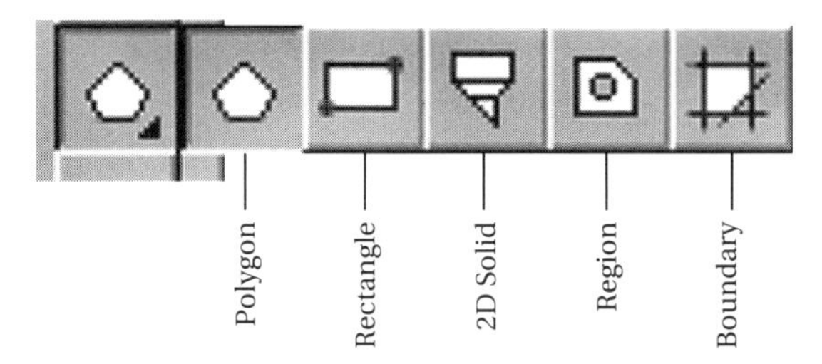

Figure 1.70 Select an icon with a triangle from the `Draw` toolbar to display flyouts, including Polygon.

ELLIPSE (DRAW TOOLBAR)

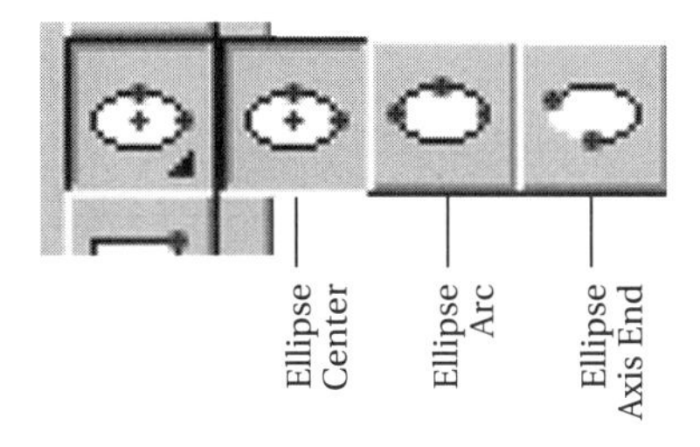

Figure 1.71 Select an icon with a triangle from the `Draw` toolbar to display Ellipse flyouts.

When the `EDGE` option is used, the next prompt asks:

`First endpoint of edge:` (Select point.)

`Second endpoint of edge:` (Select point.)

A polygon is drawn in a counterclockwise direction about the center point. Polygons can have a maximum of 1024 sides.

1.20 Ellipse (Draw Toolbar)

The `ELLIPSE` icon in the `Draw` toolbar gives flyouts for 3 types of ellipses **(Fig. 1.71)**. An ellipse is drawn by selecting the endpoints of the major axis and a third point, `P3`, to give the length of the minor radius **(Fig. 1.72)**.

In **Fig. 1.73**, points are picked at the center of the ellipse at `P1`, one axis endpoint at `P2`, and the second axis length from `P1` to `P3`. The ellipse is drawn through points `P2` and the endpoint of the minor diameter specified by `P3`.

The endpoints of the axis can be located and the rotation angle specified. An angle of 0° gives a full circle, and an angle of 90° gives an edge.

ELLIPSE: ENDPOINTS (DRAW TOOLBAR)

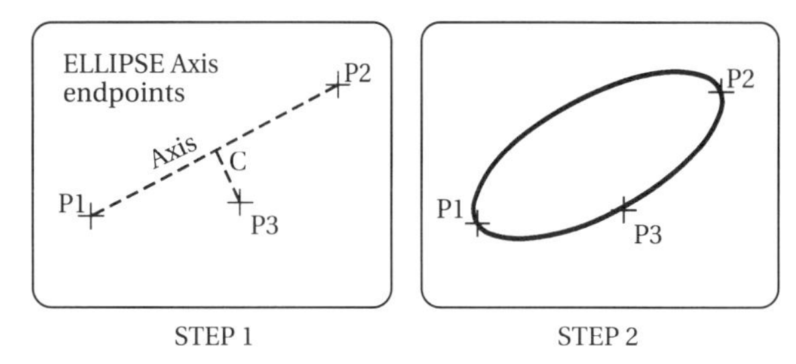

Figure 1.72 `ELLIPSE`: Endpoints option.
Step 1 `Command:` <u>`ELLIPSE`</u> (ENTER)
`Arc/Center/<Axis endpoint 1>` <u>`P1`</u>
`Axis endpoint 2:` <u>`P2`</u>
`<Other axis distance>/Rotation:` <u>`P3`</u>
Step 2 The ellipse is drawn through <u>`P1`</u> and <u>`P2`</u> and a point at the distance `C-P3` measured perpendicularly from `P1-P2`.

ELLIPSE: CENTER & AXIS (DRAW TOOLBAR)

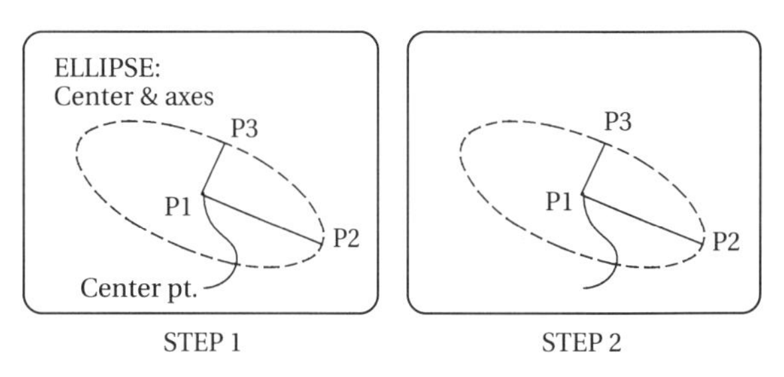

Figure 1.73 `ELLIPSE`: Center and axes.
Step 1 `Command:` <u>`ELLIPSE`</u> (ENTER)
`(Axis endpoint 1)/Center:` <u>`C`</u> (ENTER)
`Center of ellipse: P1`
`Axis endpoint: P2`
Step 2 `<Other axis distance>/Rotation:` <u>`P3`</u>
(The ellipse is drawn.)

EFFECTS OF FILL ON AND OFF

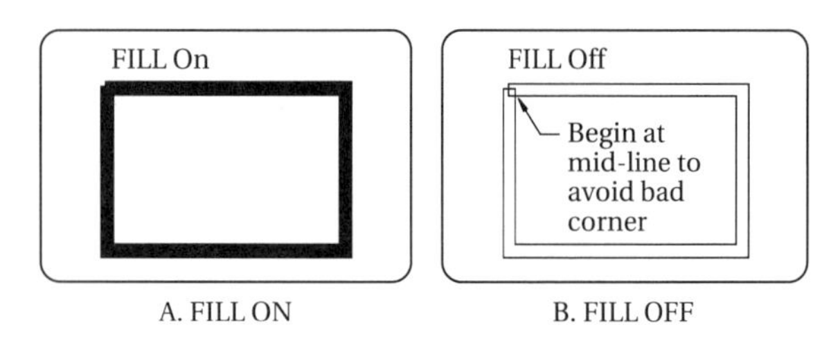

Figure 1.74 `TRACE` command.
A The `TRACE` command is used with `Fill On`, solid lines are drawn to the width specified.
B When `Fill` is turned `Off`, parallel lines are drawn.

STANDARD TOOLBAR: PART 3

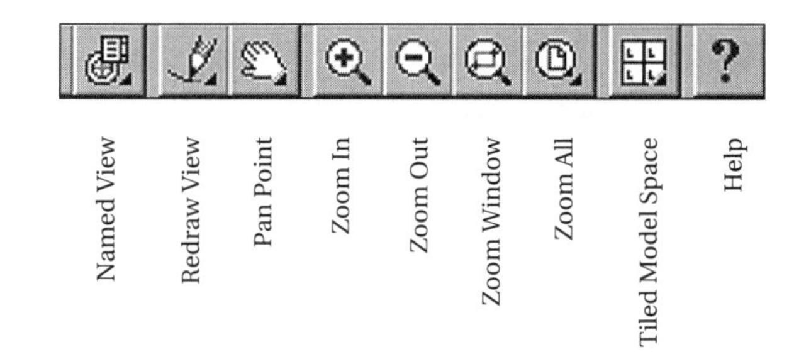

Figure 1.75 This part of the `Standard` toolbar contains many options for `PAN`ning and `ZOOM`ing views of the screen.

1.21 Trace (Miscellaneous Toolbar)

Wide lines made with multiple strokes can be drawn with the `TRACE` command as follows:

```
Command: TRACE (ENTER)
Width: 0.4 (ENTER)
From point: Select points (ENTER)
```

Type `FILL` and set to `On` and the `TRACE` will be drawn as shown in **Fig. 1.74A**. When `FILL` is `Off`, the lines will be drawn as parallel lines with "mitered" angles **(Fig. 1.74B)**. Under `Options` and `Display`, `Solid Fill` is checked for `On`.

1.22 Zoom and Pan (Standard Toolbar)

Parts of a drawing can be enlarged or reduced by the `ZOOM` command in the submenu under the `VIEW` pull-down menu or the `Standard` toolbar **(Fig. 1.75)**. **Figure 1.76** shows how a `ZOOM` window is used to enlarge part of a drawing.

`ZOOM (Z)` has the following options: `All`, `Center`, `Dynamic`, `Extents`, `Left`, `Previous`, `Vmax`, or `Scale(X/XP)`.

`All` expands the drawing's `Limits` (Grid) to fill the screen.

`Previous` displays the last `ZOOM`ed view.

`Scale X/XP` magnifies a drawing relative to its current size or to paper space. By typing `1/4X` or `.25X`, the drawing is scaled so that it is displayed at one-fourth of its previous size. By typing `1/4XP` or `.25XP`, the drawing will be scaled so that $\frac{1}{4}$ inch equals 1 inch.

`Dynamic` lets you `ZOOM` and `PAN` by selecting points with the cursor.

`Center` is picked to specify the center of the `ZOOM`ed image and specify its magnification or reduction.

ZOOM COMMAND (STANDARD TOOLBAR)

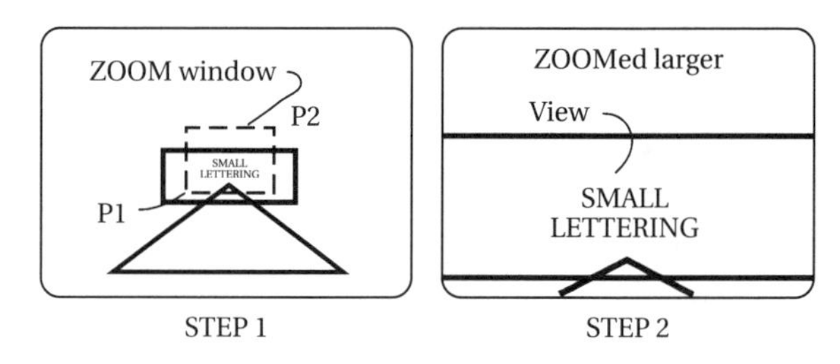

Figure 1.76 `ZOOM` command.
Step 1 `Command: ZOOM` (ENTER)
`All/Center/Dynamic/Extents/Left/Previous/Vmax/Window/<Scale (X/XP)>: W` (ENTER)
`First corner: P1`
Step 2 `Other corner: P2` (The window is enlarged to fill the screen.)

PAN COMMAND (STANDARD TOOLBAR)

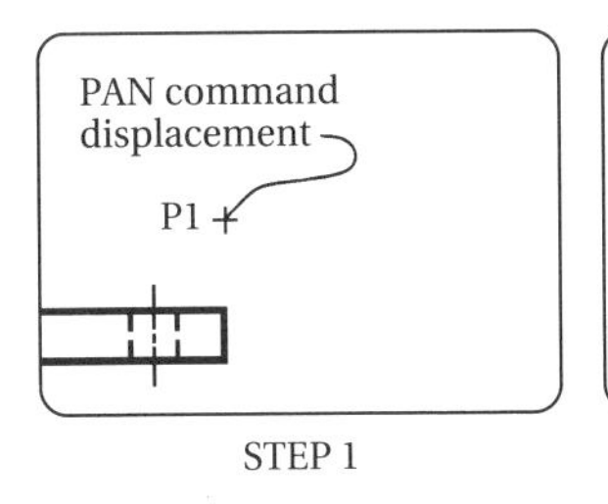

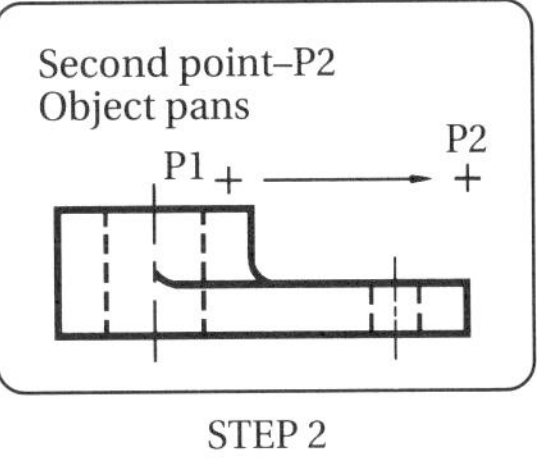

Figure 1.77 PAN command.
Step 1 Command: PAN (ENTER)
PAN Displacement: P1
Step 2 Second point: P2 (The viewpoint is moved to new position.)

Left establishes the lower-left corner and the height of the drawing to be ZOOMed.

Extents enlarges the drawing to its maximum size on the screen.

Vmax displays the current screen images as large as possible without forcing a complete regeneration.

The PAN (P) command **(Fig. 1.77)** is used to pan the view across the screen by selecting two points. The drawing is not relocated; only your view of it is changed.

SELECTION OF OBJECTS

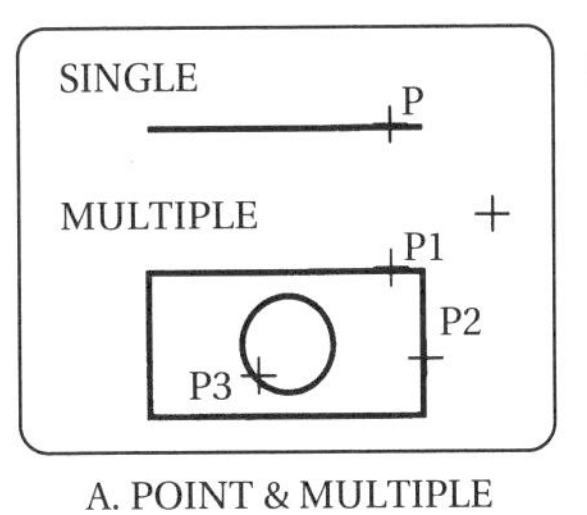

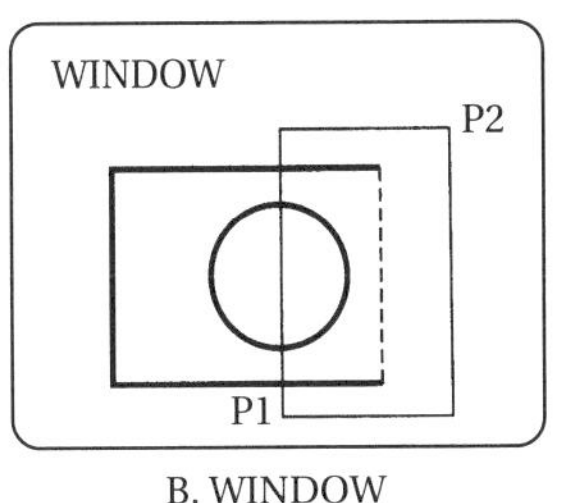

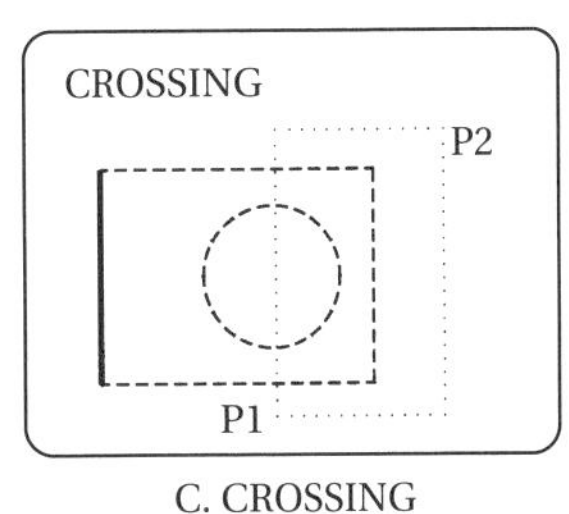

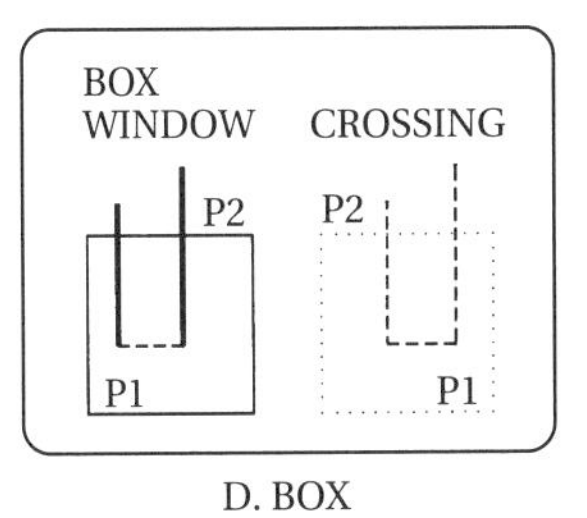

Figure 1.78 Select options.
A When prompted by the command to Select objects:, objects can be selected individually.
B A Window (W) can be used to select objects lying completely within the window.
C The Crossing Windows (C) option can be used to select objects within or crossed by the window.
D The Box option is used for both the window and crossing options, determined by the sequence in which the diagonal corners are selected.

1.23 Selecting Objects

A reoccurring prompt, Select objects:, asks you to select an object or objects that are to be ERASEd, CHANGEd, or modified in some way. Type SELECT from the current command that requires a select (MOVE, for example) and the select options will be displayed: Auto, Add, All, Box, Crossing, CPolygon, Fence, Group, Last, Multiple, Previous, Remove, Single, Undo, Window, WPolygon.

Figure 1.78 shows ways of selecting objects for applicable commands such as MOVE, COPY, ERASE, and so forth when prompted to Select objects. The application of the options of Single and Multiple, Window, Crossing, and Box are shown in parts A, B, C, and D, respectively.

Last is used to pick the most recently drawn object.

Previous recalls the previously selected set of objects. For example, enter MOVE, and type P, and the last selected objects are recalled.

Undo removes objects in reverse order one at a time by typing U (undo) repetitively.

Remove is used while selecting objects to switch from adding objects to the selection set (Select objects:) to removing objects (Remove objects:). Switch to remove mode by typing R. Each object selected is then removed from the set of selected objects. To switch back to select (add) mode, type A. You

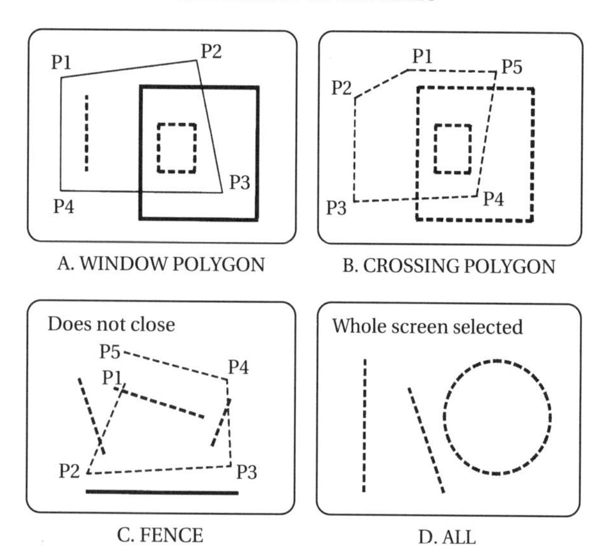

Figure 1.79
A A `Window Polygon` (`WP`) selects objects lying within it.
B A `Crossing Polygon` (`CP`) selects objects within it and crossed by it.
C A `Fence` (`F`) is a nonclosing polyline that selects entities that it crosses.
D `All` selects everything on the screen.

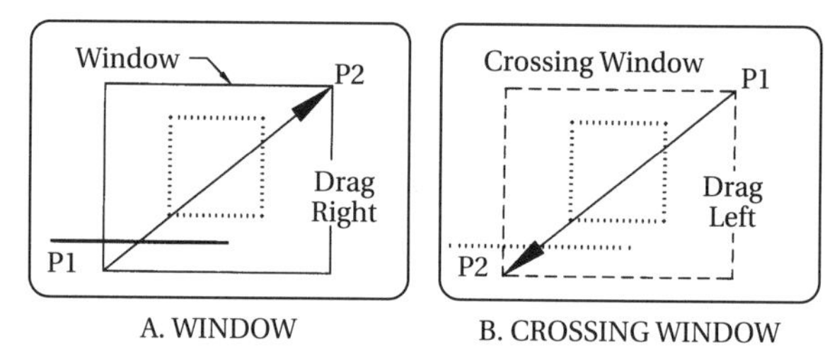

Figure 1.80 Selection windows.
A A window is formed by pressing and holding the select button while dragging a diagonal corner to the `right`.
B A crossing window is formed in the same manner, but it is dragged to the `left`.

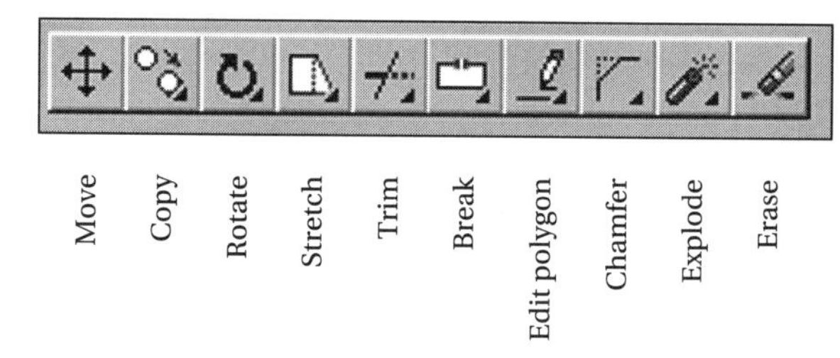

Figure 1.81 The `Modify` toolbar has ten options for making changes to drawings.

can also remove a previously selected object from the set by holding down the (SHIFT) key as you select it.

`Single` (`SI`) causes the program to act on the object or sets of objects without pausing for a response.

`WPolygon` (`WP`) forms a solid-line polygon that has the same effect as a window **(Fig. 1.79)**.

`CPolygon` (`CP`) forms a dotted-line polygon that has the same effect as a crossing window (Fig. 1.79).

`Fence` selects corner points of a polyline that will select any object it crosses the same as a crossing window (Fig. 1.79).

`All` selects everything on the screen (Fig. 1.79).

A `Window` or a `Crossing window` can be obtained automatically by pressing the pick button, holding it down, and selecting the diagonal of a window. By dragging it to the right, a window is obtained; by dragging it to the left, a crossing window is obtained **(Fig. 1.80)**.

1.24 ERASE and BREAK (Modify Toolbar)

The `ERASE` (`E`) command from the `Modify` toolbar **(Fig. 1.81)** deletes specified parts of a drawing. The selection techniques described previously can be used to select objects to be erased as shown in **Figs. 1.82** and **1.83**.

The default of the `ERASE` command, `Select Objects`, allows you to pick one or more objects

ERASING: W OPTION (MODIFY TOOLBAR)

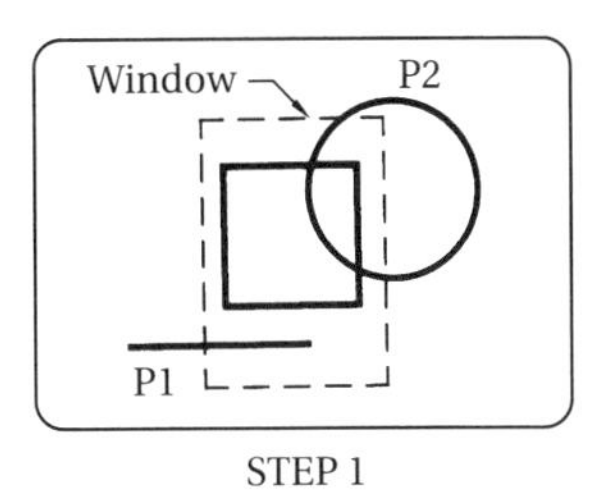

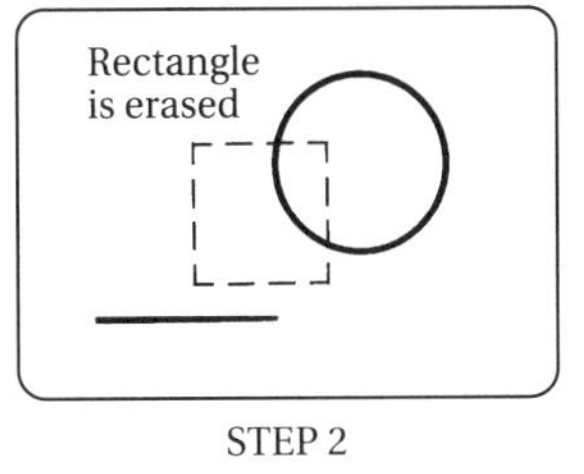

Figure 1.82 ERASE: Window option.
Step 1 Command: ERASE (ENTER)
Select objects: WINDOW or W (ENTER)
First corner: P1
Other corner: P2
Step 2 Select objects: (ENTER) (Entities entirely within window are erased.)

BREAK COMMAND (MODIFY TOOLBAR)

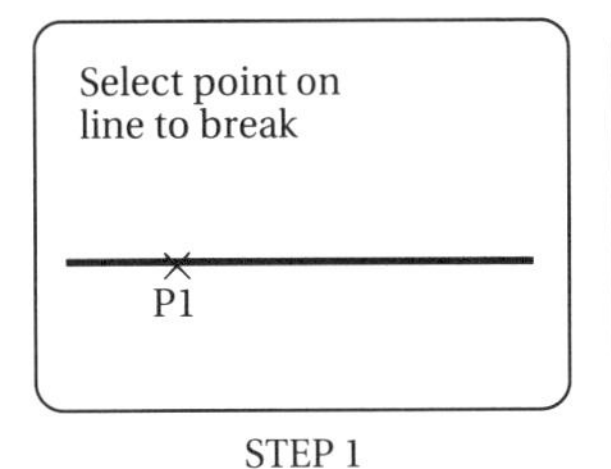

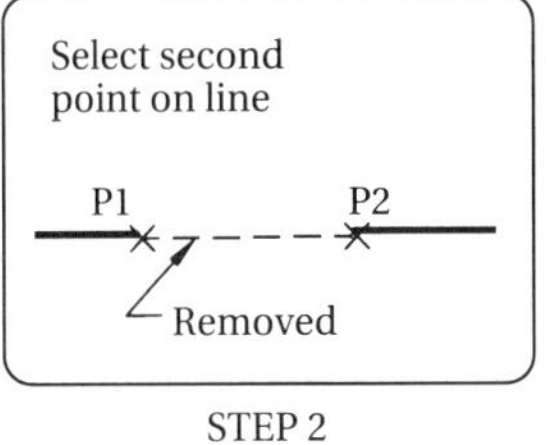

Figure 1.84 BREAK command.
Step 1 Command: BREAK (ENTER)
Select object: P1 (point of break)
Step 2 Enter second point (or F for first point): P2 (The line is removed from P1 to P2.)

ERASING: CROSSING WINDOW (MODIFY TOOLBAR)

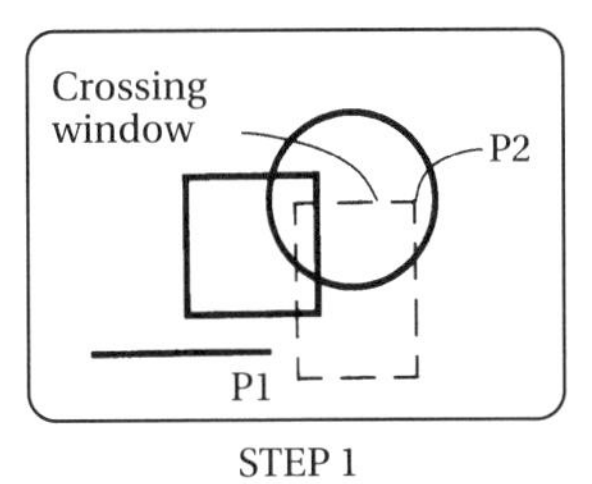

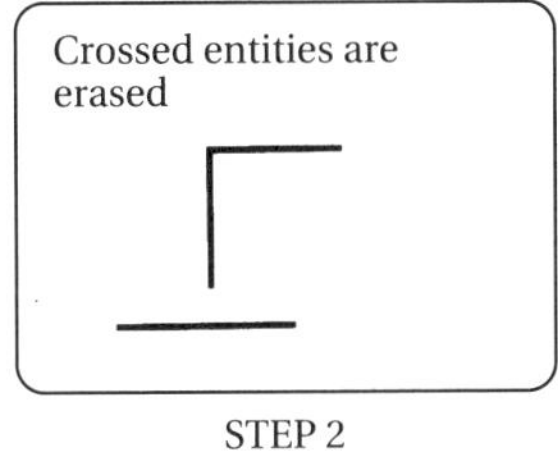

Figure 1.83 ERASE: Crossing option.
Step 1 Command: ERASE (ENTER)
Select objects: C (Crossing) (ENTER)
First corner: P1
Other corner: P2
Step 2 Select objects: (ENTER) (Entities crossed by or in window are erased.)

BREAK COMMAND

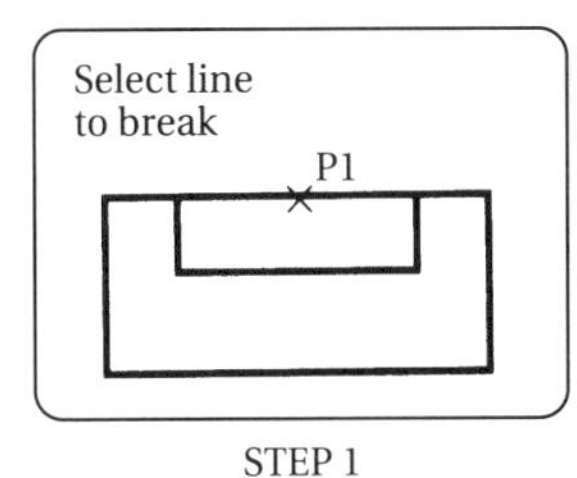

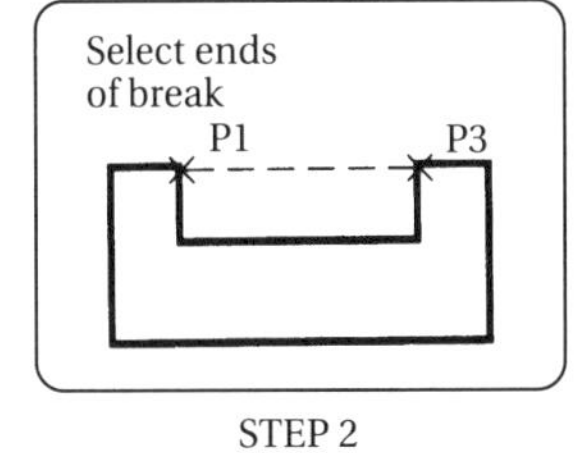

Figure 1.85 BREAK F option.
Step 1 Command: BREAK (ENTER)
Select object: P1 (on line to be broken)
Enter second point (or F for first point): F
Step 2 Enter first point: P2
Enter second point: P3 (The line is broken from P2 to P3.)

and delete them by pressing (ENTER). Type OOPS to restore the last erasure, but only the last one.

The BREAK command removes part of a line, pline, arc, or circle **(Fig. 1.84)**. To specify a break at an intersection with another line as shown in **Fig. 1.85,** select the line to be broken and select the F option. The endpoints can be selected without fear of selecting the wrong lines.

1.25 MOVE and COPY (Modify Toolbar)

The MOVE (M) command repositions an object **(Fig. 1.86)** and the COPY command duplicates it, leaving the original in its original position. The COPY (C) command is applied in the same manner as the MOVE command. The COPY command has a Multiple option for locating multiple copies of drawings in different positions.

MOVE COMMAND

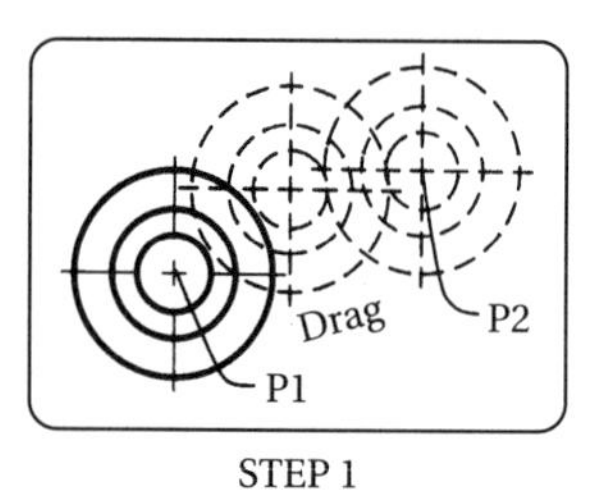

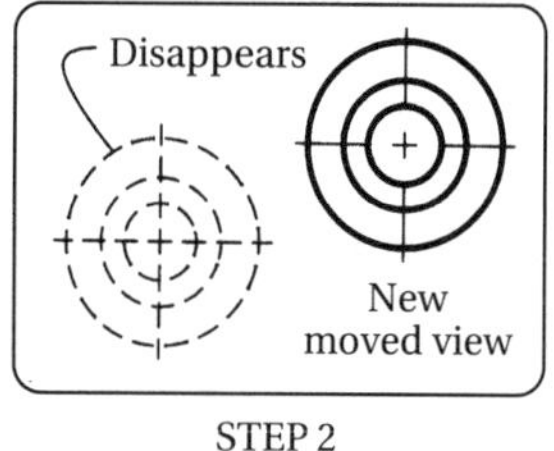

Figure 1.86 `MOVE` command.
Step 1 `Command:` <u>`MOVE`</u> (ENTER)
`Select objects:` <u>`W`</u> (ENTER) (window object)
`Base point or displacement:` <u>`DRAG`</u> (ENTER)
`Base point or displacement:` <u>`P1`</u> (or X,Y distance)
`Second point of displacement:` <u>`P2`</u> (The drawing is `DRAG`ged as the cursor is moved.)
Step 2 When moved to `P2`, press the `Select` button to draw the circle. If `DRAGMODE` or `AUTO` is `On`, dragging is automatic.

1.26 TRIM (Modify Toolbar)

The `TRIM` command selects cutting edges for trimming selected lines, arcs, or circles that cross the edges at their crossing points **(Fig. 1.87)**. The `Crossing` option of `TRIM` is used to select four cutting edges in **Fig. 1.88** and the lines are trimmed.

1.27 EXTEND (Modify Toolbar)

The `EXTEND` command lengthens lines, plines, and arcs to intersect a selected boundary **(Fig. 1.89)**. You are prompted to select the boundary object and the object to be extended, which will extend the object to the boundary. More than one object can be extended. `EXTEND` will not work on "closed" `Plines` such as polygons.

1.28 UNDO (Standard Toolbar)

The `UNDO (U)` command can reverse the previous commands one at a time back to the beginning of a session. The `REDO` command reverses the last `UNDO`; `OOPS` will not work. The `UNDO` command has these options: `Auto`, `Control`, `BEgin`, `End`, `Mark`, `Back`, and `Number`.

TRIM COMMAND

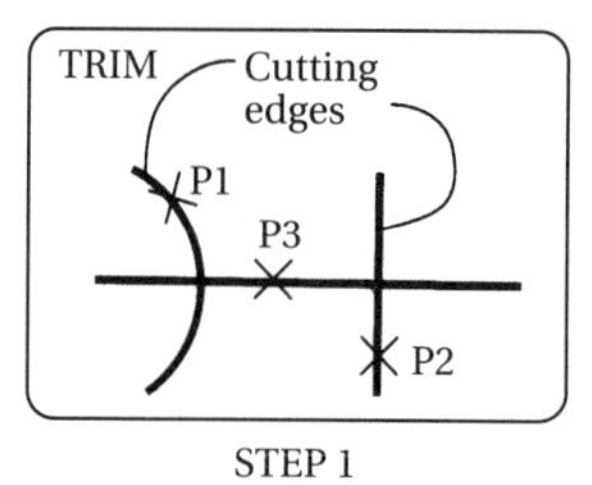

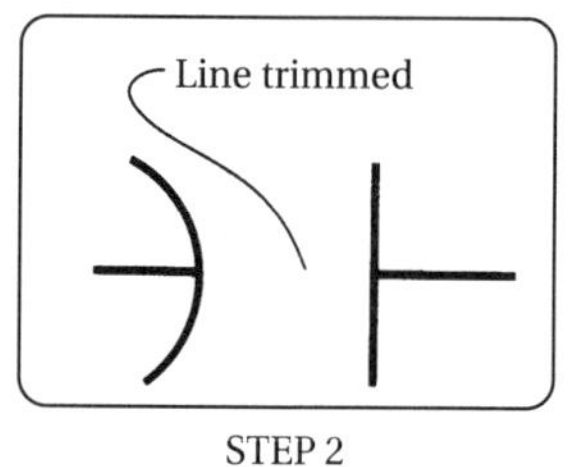

Figure 1.87 `TRIM` edges.
Step 1 `Command:` <u>`TRIM`</u> (ENTER)
`Select cutting edge:` (Projmode=UCS, Edgemode= No extend)
`Select objects:` <u>`P1`</u>
`Select objects:` <u>`P2`</u>
`Select objects:` (ENTER)
Step 2 `<Select object to trim>/Project/Edge/ Undo:` <u>`P3`</u> (segment between cutting edges is removed.)

TRIM: CROSSING

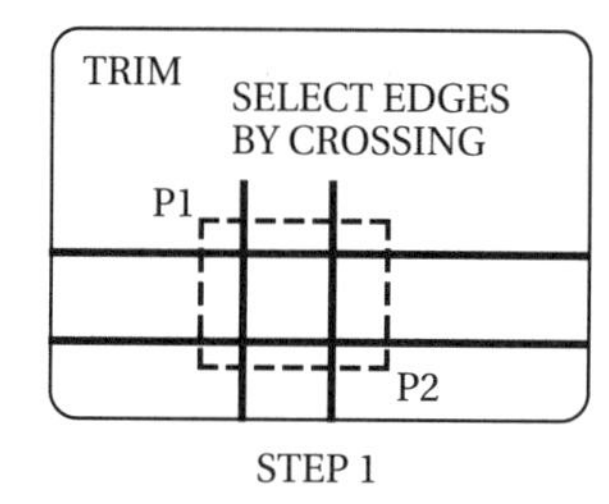

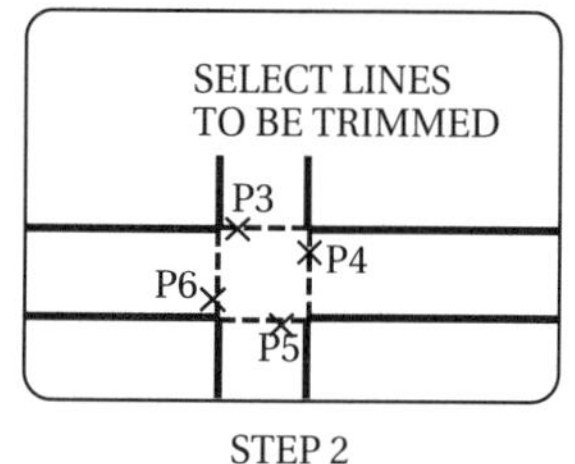

Figure 1.88 `TRIM` using crossing window.
Step 1 `Command:` <u>`TRIM`</u> (ENTER)
`Select cutting edge:` (Projmode=UCS, Edgemode= No extend)
`Select objects:` <u>`C`</u>
`First corner:` <u>`P1`</u>
`Other corner:` <u>`P2`</u>
`Select objects:` (ENTER)
Step 2 `<Select object to trim>/Project/Edge/ Undo:` <u>`P3, P4, P5, P6`</u> (The line segments are trimmed one at a time.)

`Command:` <u>`UNDO`</u> (ENTER)

`Auto/Cntrol/BEgin/End/Mark/Back/ <Number>:` <u>`4`</u> (ENTER)

EXTEND COMMAND (MODIFY TOOLBAR)

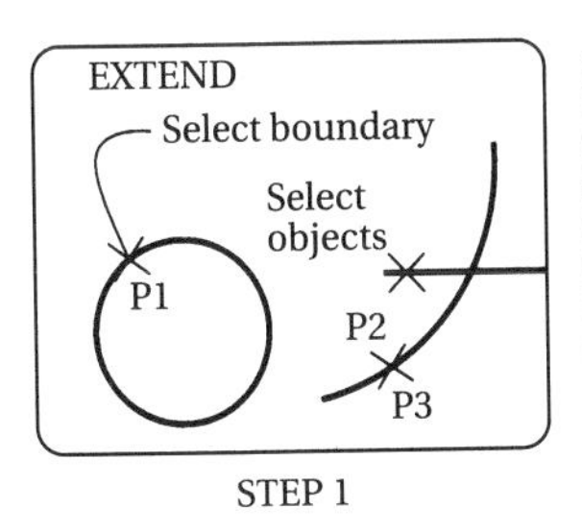

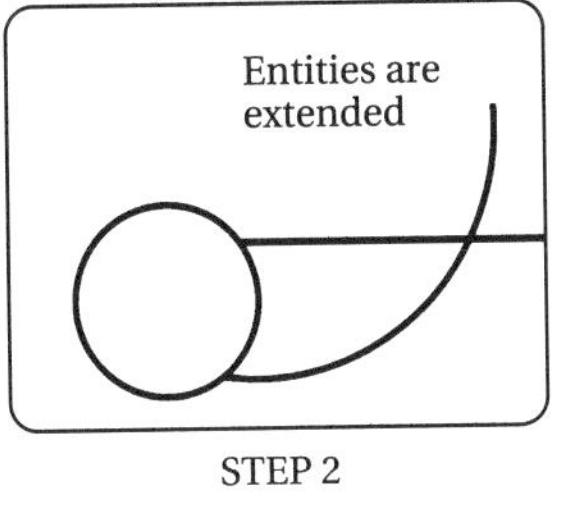

Figure 1.89 `EXTEND` command.
Step 1 `Command:` EXTEND (ENTER)
`Select cutting edge: (Projmode=UCS, Edgemode = No extend)`
`Select objects:` P1 (ENTER)
Step 2 `<Select object to extend>..` P2 and P3 (The line and arc are extended to the boundary.)

CHANGE A LINE (COMMAND TOOLBAR)

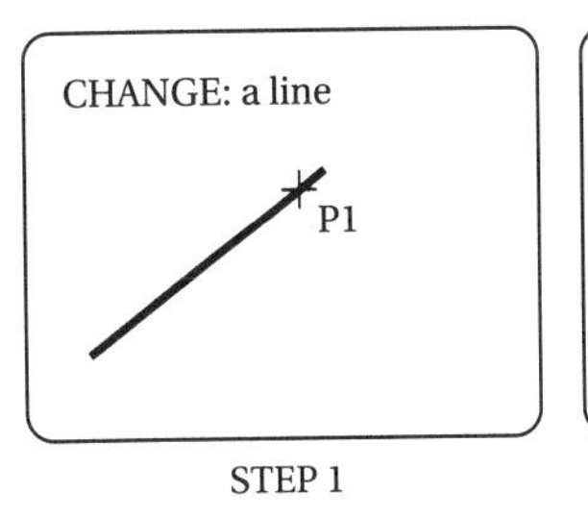

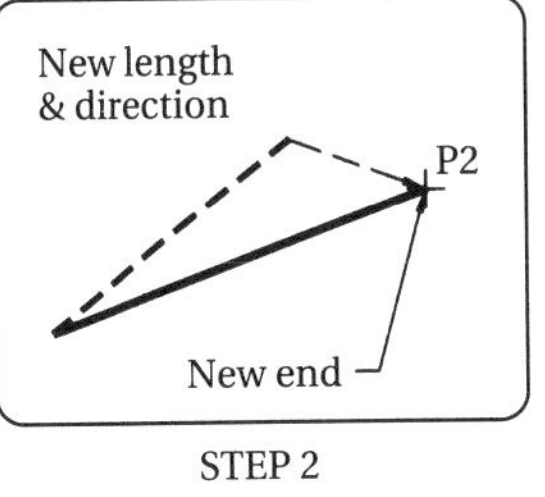

Figure 1.90 `CHANGE` Command: Line.
Step 1 `Command:` CHANGE (ENTER)
`Select objects:` P1
`Select objects:` (ENTER)
Step 2 `Properties/<Change point>:` P2 (ENTER)
(End of line is moved.)

CHANGE A CIRCLE

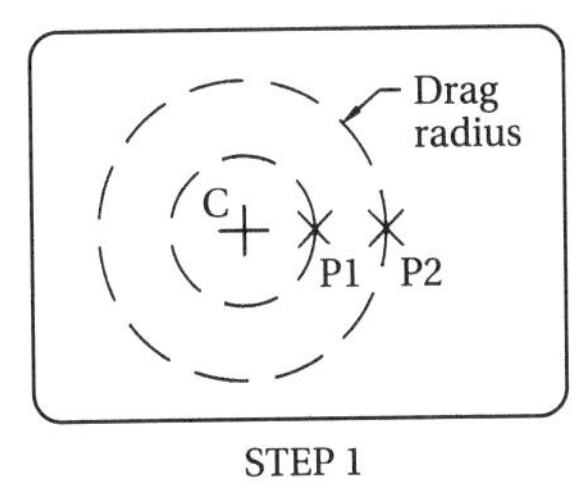

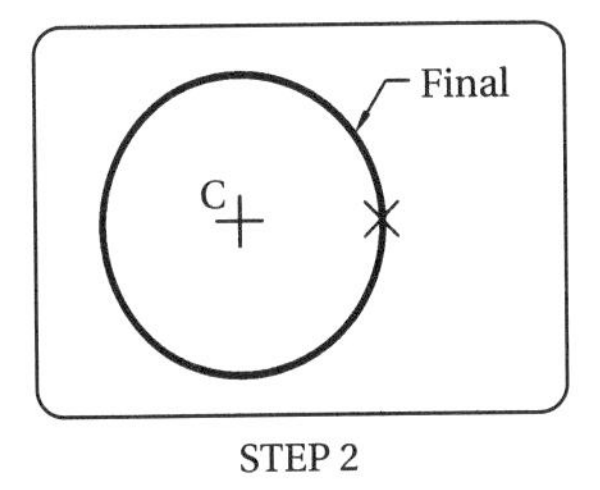

Figure 1.91 `CHANGE` Command: Circle.
Step 1 `Command:` CHANGE (ENTER)
`Select objects:` P1 (ENTER)
Step 2 `Properties/<Change point>:` P2 (To change radius.)

Entering 4 has the same effect as using the `U` command four separate times.

`Mark` identifies a point in the drawing process to which subsequent additions can be undone by the `Back` option. Only the part of the drawing added after placing the `Mark` will be undone. If you use the `Back` option and no previous `Mark` has been left, AutoCAD prompts:

```
This will undo everything.
OK? <Y>. Y (ENTER)
```

By responding `Y`, everything you've added to the drawing since it was loaded will be removed.

The `BEgin` option groups a sequence of operations until `End` option terminates the group. `UNDO` treats the group as a single operation.

The `CONTROL` subcommand has three options: `All`, `None`, and `One`. `All` turns on the full features of the `UNDO` command, `None` turns them off, and `One` uses `UNDO` commands for single operations and requires the least disk space.

1.29 CHANGE (Modify Toolbar)

Type `CHANGE` to modify features: `Lines`, `Circles`, `Text`, `Attribute Definitions`, `Blocks`, `Color`, `Layers`, `Linetypes`, and `Thickness`. The position of an endpoint of a `Line` is changed by selecting one end and locating a new endpoint **(Fig. 1.90)**. `CHANGE` varies the size of a circle by picking a point on its arc and dragging to a new size **(Fig. 1.91)**.

The `Text` option of `CHANGE` modifies text by pressing (ENTER) until the prompts `Style`, `Height`, `Rotation Angle`, and `Text String` appear in sequence **(Fig. 1.92)**. `Attribute Definitions`, including `Tag`, `Prompt String`, and `Default Value`, can be revised with the `CHANGE` command. `BLOCK`s can be moved or rotated with the `CHANGE` command as shown in **Fig. 1.93**.

CHANGE TEXT

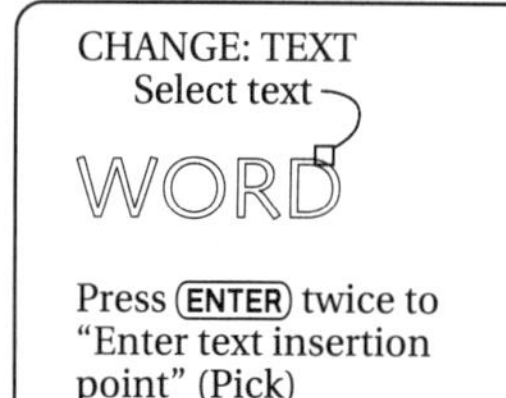

STEP 1

Text style: RT
Text height: 0.2

WORD
words

Rotation angle: 0
Text string: WORDS

STEP 2

Figure 1.92 CHANGE text.
Step 1 `Command:` <u>CHANGE</u> ENTER
`Select objects:` (Select text.)
`Select objects:` ENTER
`Properties/<Change point>` ENTER
`Enter text insertion point:` (Select point.)
Step 2 `Text Style:` <u>Standard</u>
`New style or RETURN for no change:` <u>RT</u>
`New height <.50>:` <u>0.20</u>
`New rotation angle <0>:` ENTER
`New text <WORD>:` ENTER

CHANGE A BLOCK

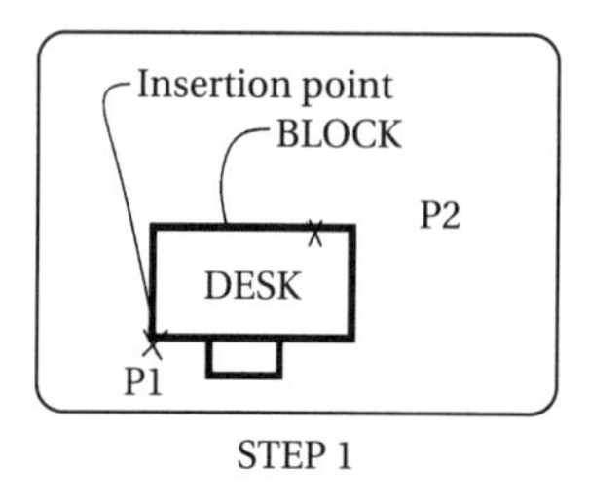

STEP 1

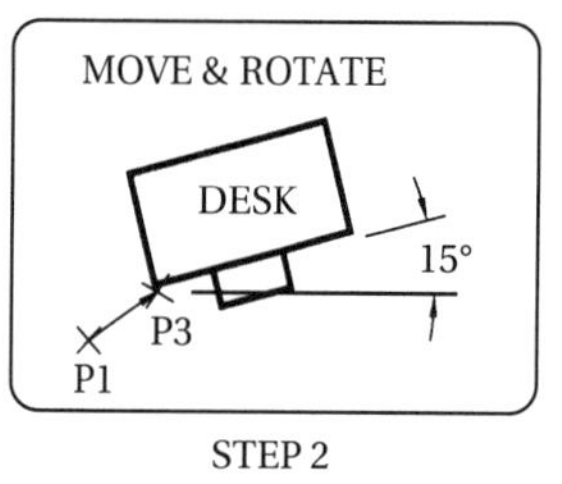

STEP 2

Figure 1.93 CHANGE Command: Block.
Step 1 `Command:` <u>CHANGE</u> ENTER
`Select objects:` <u>P1</u>
`Select objects:` ENTER
`Properties/<Change point>:` ENTER
`Change point (or Layer or Elevation):` <u>P2</u>
(Select new position.)
Step 2 `Enter block insertion point:` <u>P3</u>
`New rotation angle <0>:` <u>15</u> ENTER (The block is moved and rotated 15 degrees.)

Property changes of the CHANGE command, LAyer, Color, LType, and Thickness, are made by selecting objects and typing P (properties) as shown in **Fig. 1.94**. Type LA (layer) and the name

CHANGE LAYERS

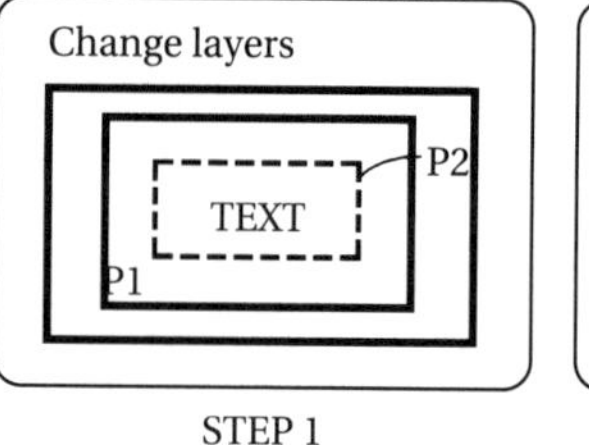

STEP 1

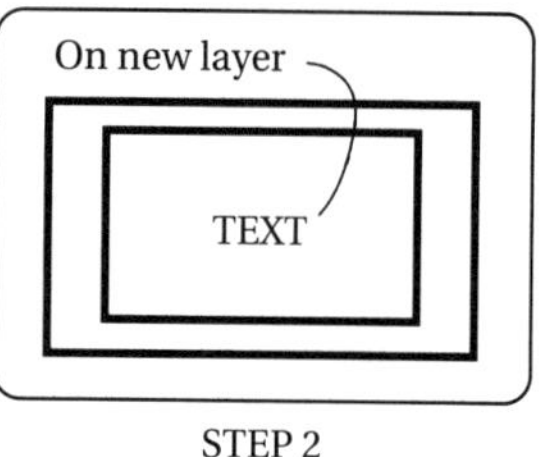

STEP 2

Figure 1.94 CHANGE Command: Layers.
Step 1 `Command:` <u>CHANGE</u> ENTER)
`Select objects:` <u>W</u> ENTER
`First corner:` <u>P1</u> ENTER
`Other corner:` <u>P2</u> ENTER
Step 2 `Properties < Change point >:` <u>P</u>
`Change what property (Color/Elev/LAyer/LType/ltScale/Thickness)?` <u>LA</u> ENTER
`New layer:` <u>VISIBLE</u> (An existing layer.) ENTER

of the layer on which the text is to be changed.

Multiple Colors and LTypes can be assigned to objects on the same layer by the CHANGE command, but it is better for each layer to have only one layer and linetype. The Thickness property is the height of an extrusion in the Z-direction of a three-dimensional object drawn with the Elev and Thickness options. Changing the Thickness of a 2D surface from 0 to a nonzero value converts it to an extruded 3D drawing.

At the Command line, type DDMODIFY and select an object when prompted to display a Modify dialog box specific to the type of object selected. You can then change any of the properties (color, layer, linetype, thickness, size, and so on) of the selected object **(Fig. 1.95)**.

1.30 CHPROP (Command Line)

Type CHPROP at the Command line to change object properties (Color, LAyer, LType, ltScale, and Thickness) regardless of their extrusion direction. This is a useful option when modifying 3D drawings. CHPROP is used as follows:

`Command:` <u>CHPROP</u>

MODIFY LINE DIALOG BOX (DDMODIFY)

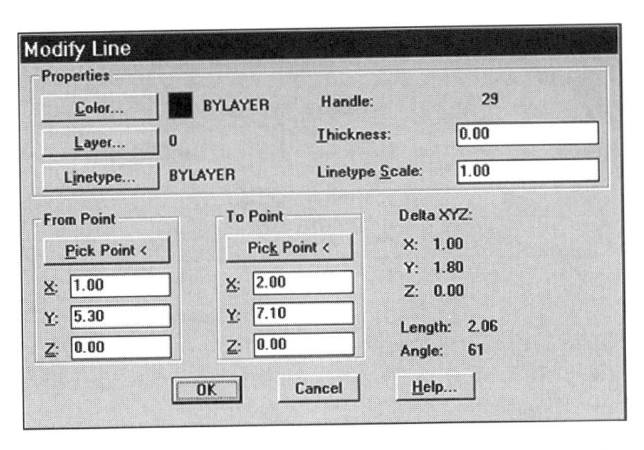

Figure 1.95 The `Modify` (`DDMODIFY`) box can be used to modify all objects in the same manner as when the `CHANGE` command is used. Type `DDMODIFY` at the `Command` line and select the object to be modified to obtain a `Modify` box.

`Select objects:` (Select objects.)

`Change what property (Color/LAyer/LType/ltScale/Thickness)?`

Make the same choice as when using the `CHANGE` command.

1.31 GRIPS (Options Menu)

Grips are small squares that appear on selected objects at midpoints and ends of lines, at centers and quadrant points of circles, and at insertion points of text. `Grips` are used to `STRETCH`, `MOVE`, `ROTATE`, `SCALE`, and `MIRROR`.

The `Grips` dialog box (`DDGRIPS`) is found under `Options` of the menu bar **(Fig. 1.96)**. The `Enable Grips` check box turns on grips for all objects. `Enable Grips Within Blocks` turns on grips for objects within a `Block`; when off, a single grip is given at the insertion point of the `Block`.

`Grip Colors` turns on the `Color` dialog box for assigning colors to selected and unselected grips; unselected grips are not filled in. `Grip Size` sets the size of grip boxes with a slider box.

GRIPS DIALOG BOX (DDGRIPS)

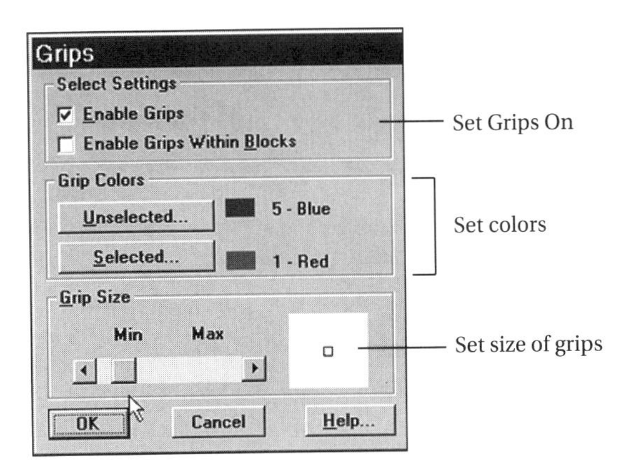

Figure 1.96 From the `Options` menu select `Grips` to obtain this box, which allows you to enable or disable grips, assign grip colors, and set grip sizes.

GRIPS: STRETCH

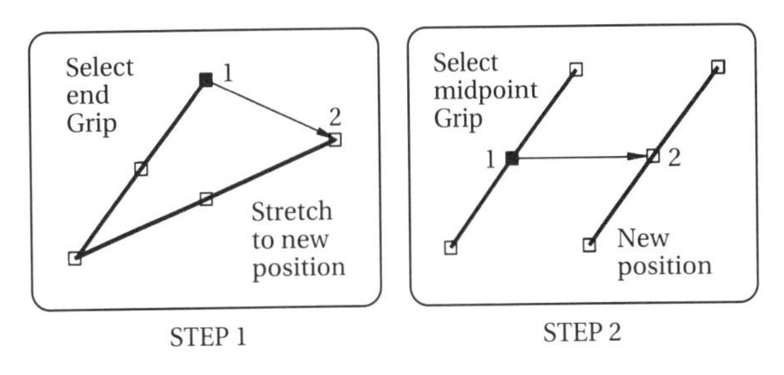

Figure 1.97 Stretching with `Grips`.
Step 1 Select the line and grips appear; click on the end-point (`P1`) and move to the new position (`P2`).
Step 2 Select the midpoint grip (`P1`), pick a second point (`P2`), and the line is moved to it.

Using Grips By selecting an object with the cursor, grips will appear on it as open boxes. A grip that is picked and made a "hot" point is filled with color. By holding down the (SHIFT) key, more than one grip can be picked as a "hot" point, but the last grip of a series must be selected without pressing (SHIFT). Press the (ESC) key to remove grips. By turning grips on and successively pressing (ENTER), the options `STRETCH`, `MOVE`, `ROTATE`,

GRIPS: MOVING A BLOCK

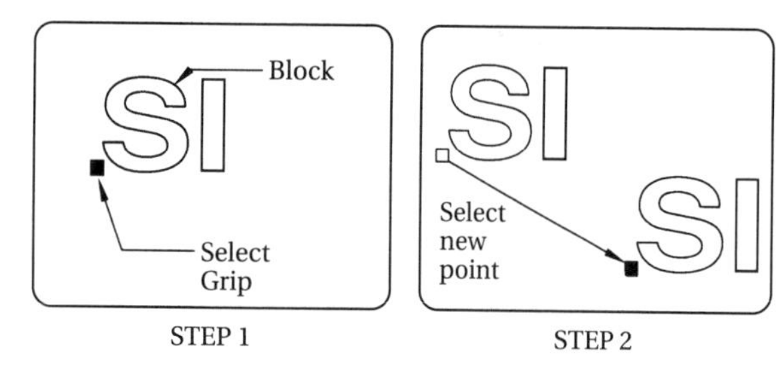

Figure 1.98 Moving with `Grips`.
Step 1 Select the `Block` and a grip appears at the insertion point. Pick the grip.
Step 2 Move the cursor to a new position and the `Block` is moved.

GRIPS: SCALE OPTION

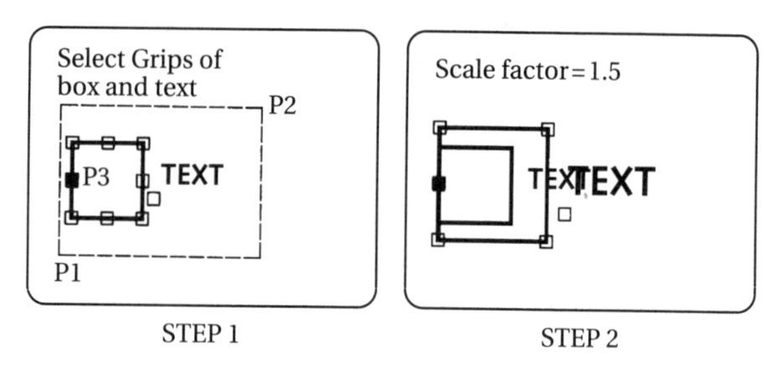

Figure 1.100 Scaling with `Grips`.
Step 1 Turn on grips by windowing; select `P3` as the base; press (ENTER) until `**SCALE**` appears.
Step 2 `<Scale factor>/Base point/Copy/Undo/Reference/eXit:` 1.5 (ENTER) (Objects are scaled.)

GRIPS: ROTATING OPTION

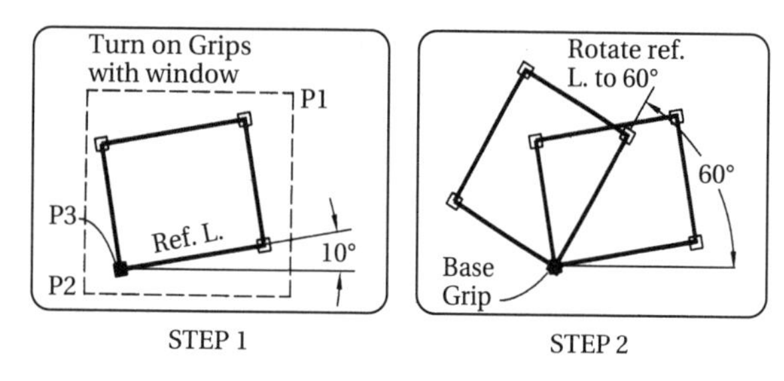

Figure 1.99 Rotation with `Grips`.
Step 1 Turn on grips by windowing; select `P3` as the base; press (ENTER) until `**ROTATE**` appears.
`<Rotation angle>/Base point/Copy/Undo/Reference/eXit:` R (ENTER)
`Reference angle <0>:` 10 (ENTER)
Step 2 `**ROTATE**`
`<New angle>/Base point/Copy/Undo/Reference/eXit:` 60 (ENTER) (Object is rotated 50° from first position.)

GRIPS: MIRROR OPTION

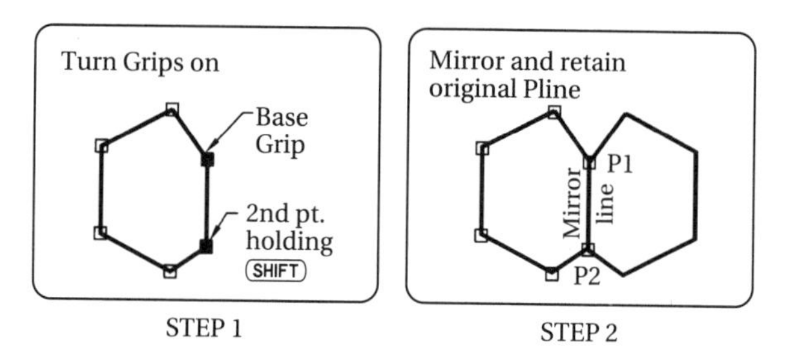

Figure 1.101 Mirroring with `Grips`.
Step 1 Turn on grips by windowing; select `P1` as the base; press (ENTER) until `**MIRROR**` appears.
Step 2 `**MIRROR**`
`<Second point>/Base point/Copy/Undo/eXit:` P2 (while holding (SHIFT) key) (ENTER) (Object is mirrored and original is retained.)

`SCALE`, and `MOVE` will be sequentially activated, each with its own subcommands.

`STRETCH` lets the endpoint grip of a line be selected as a "hot" point, and a second point as the new end of the line **(Fig. 1.97)**. Select the midpoint grip to `MOVE` the line to a new position.

To `MOVE` the `Block` in **Fig. 1.98**, select the insertion point as the hot point. Options of `Base Point`, `Copy`, `Undo`, and `eXit` can be used for these applications.

`ROTATE` revolves an object about a selected `Grip`. The `Reference` option rotates an object about a selected grip by dragging or typing a num-

POLYLINE (DRAW TOOLBAR)

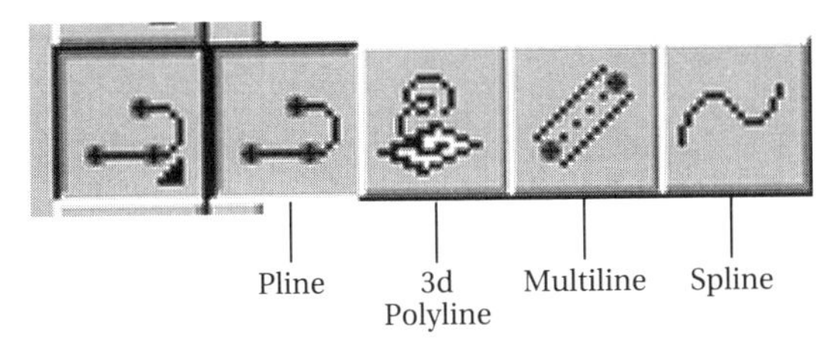

Figure 1.102 Pline is one of the flyouts you can access from the Draw toolbar.

ber. The object in **Fig. 1.99** is rotated 60° from the reference line by typing 60.

SCALE uses the grip selected as a base point to size the object **(Fig. 1.100)**. The scale factor is assigned by typing, dragging, or by selecting a reference dimension and giving it a new dimension.

MIRROR makes a mirror image of an object. The original object is removed when two grips are selected to specify a mirror line **(Fig. 1.101)**. Hold down the SHIFT key while selecting the second grip point on the mirror line and the initial object will not be removed. The mirror line need not be a line on the drawing being mirrored.

1.32 POLYLINE (Draw Toolbar)

The PLINE (PL) command from the Draw toolbar **(Fig. 1.102)** is used for drawing 2D polylines, which are lines of continuously connected segments instead of disconnected segments as drawn by the LINE command. The thickness of a PLINE can be varied as well, which requires the pen to plot with multiple strokes when plotting **(Fig. 1.103)**.

The PLINE options are Arc, Close, Halfwidth, Length, Undo, Width, and Endpoint of line. Close automatically connects the last end of the polyline with its beginning point and ends the command. Length continues a PLINE in the same direction by typing the length of the segment. If the first line was an arc, a line is drawn tangent to the arc. Undo erases the last segment of the polyline, and it can be repeated to continue erasing seg-

PLINE: WIDTH OPTION

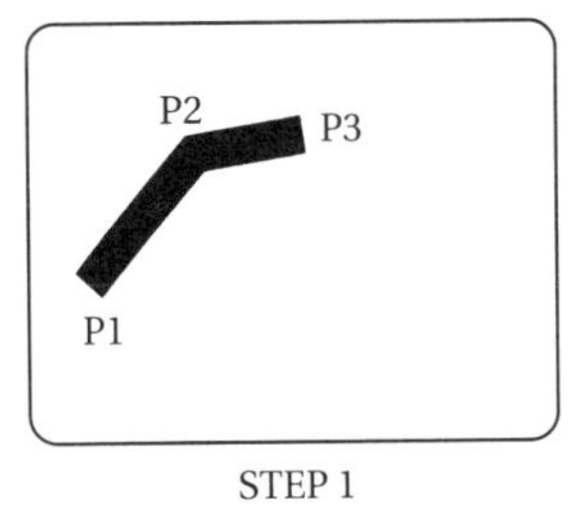

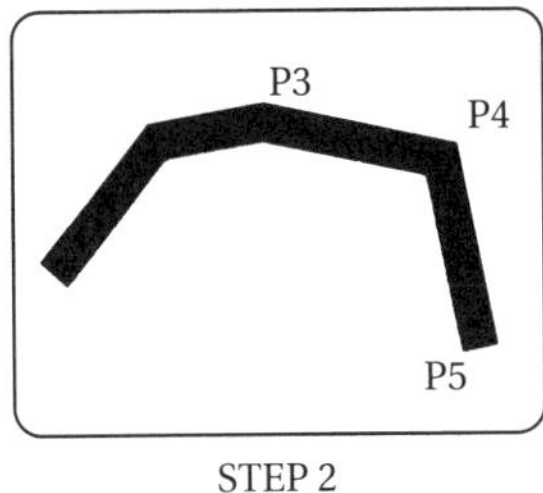

Figure 1.103 PLINE command: Width option.

Step 1 Command: PLINE ENTER
From point: P1
Current line width is 0.0000
Arc/Close/Halfwidth/Length/Undo/
Width/<Endpoint of line>: WIDTH ENTER
Starting width <0.0000>:.12 ENTER
Ending width <.12>: ENTER
Arc/Close/... /<Endpoint of line>: P2,
P2, P3
Step 2 Arc/Close/... /<Endpoint of line>:
P4, P5 ENTER

ments. Halfwidth specifies the width of the line measured on both sides of its center line.

The Arc option of PLINE is selected to obtain the prompts shown in **Fig. 1.104**. The default draws the arc tangent from the endpoint of the last line and through the next selected point. Angle gives the prompt Included angle:, to which a positive or negative value is given. The next prompt asks for Center/Radius/<End point>: to draw an arc tangent to the previous line segment. Select Center and you will be prompted for the center of the next arc segment. The next prompt is Angle/Length/<Endpoint>:, where Angle is the included angle, and Length is the chordal length of the arc. Close causes the PLINE's arc segment to close to its beginning point.

Direction lets you override the default, which draws the next arc tangent to the last PLINE segment. When prompted with Direction from starting point:, pick the beginning point and respond to the next prompt, Endpoint, by picking a second point to give the direction of the arc.

PLINE: LINES & ARCS (DRAW TOOLBAR)

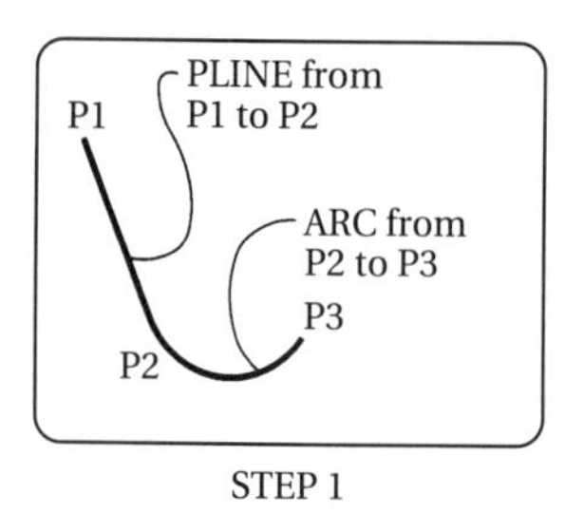

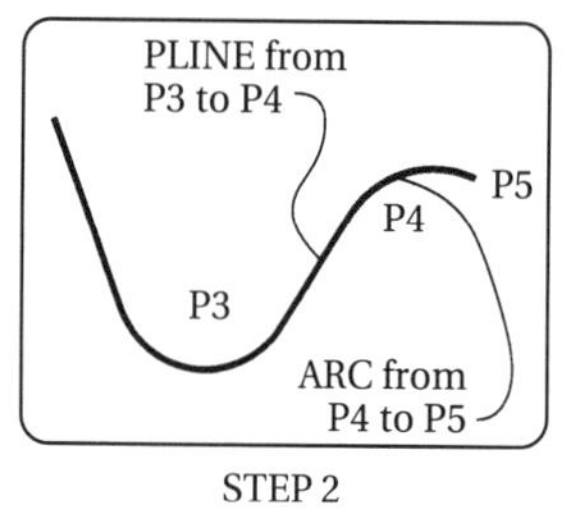

Figure 1.104 `PLINE: Lines` and `Arcs`.
Step 1 `Command:` <u>`PLINE`</u> (ENTER)
`From point:` <u>`P1`</u> (ENTER)
`Arc/Close/... /Width/<Endpoint of line>:` <u>`P2`</u> (ENTER) (ENTER)
`Command:` <u>`Arc`</u> (ENTER)
`Center/<Start point>:` <u>`P3`</u> (ENTER) (ENTER)
Step 2 `Command:` <u>`Line`</u> (ENTER)
`LINE From point:` (ENTER)
`Length of line:` <u>`P4`</u> (ENTER) (ENTER)

`Line` switches the `PLINE` command back to the straight-line mode. `Radius` gives the prompt, `Radius:`, for specifying the size of the next arc. The next prompt, `Angle/Length/<Endpoint>:`, lets you specify the included angle or the arc's chordal length. `Second Pt` gives two prompts, `Second point:` and `Endpoint:`, for selecting points on an arc.

1.33 PEDIT (Modify Toolbar)

The `PEDIT` command modifies `PLINES` with the following options: `Close`, `Join`, `Width`, `Edit vertex`, `Fit`, `Spline`, `Decurve`, `Ltype generate`, `Undo`, `eXit`. If the `PLINE` is already closed, the `Close` command will be replaced by the `Open` option.

`Join (J)` gives the prompt, `Select objects Window or Last:`, for selecting segments to join into a polyline. Segments must have exact meeting points to be joined.

`Width (W)` gives the prompt, `Enter new width for all segments:`, for assigning a new width to a `PLINE`.

PEDIT: FIT CURVE (MODIFY TOOLBAR)

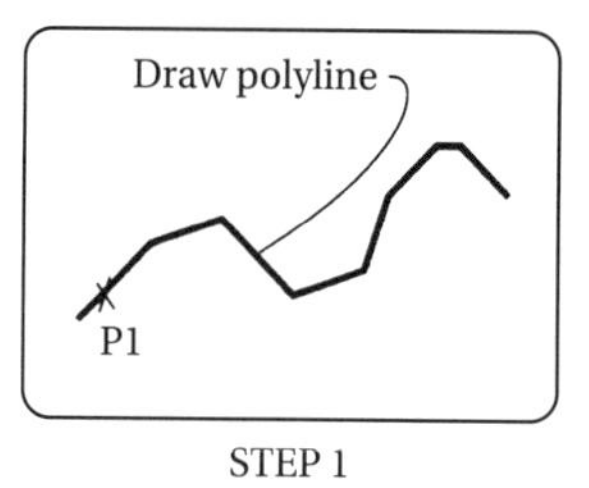

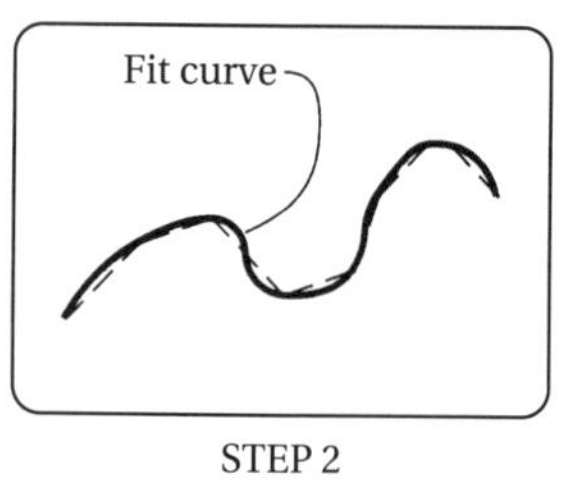

Figure 1.105 `PEDIT: Fit` curve.
Step 1 `Command:` <u>`PEDIT`</u> (ENTER)
`Select Polyline:` <u>`P1`</u>
Step 2 `Close/Join/Width/Edit vertex/Fit/Spline/Decurve/Ltype gen/Undo/eXit <X>:` <u>`Fit`</u> (ENTER) (The curve is smoothed with arcs.)

PEDIT: FIT & SPLINE

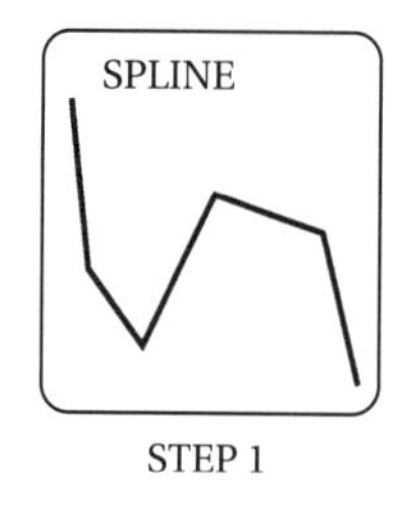

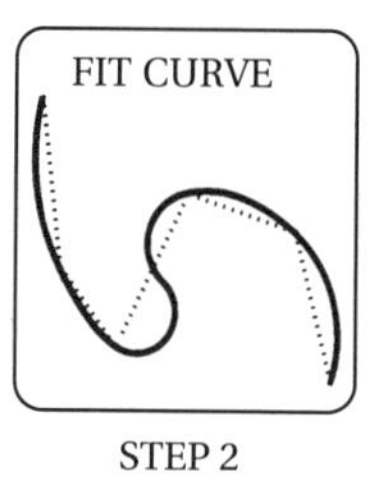

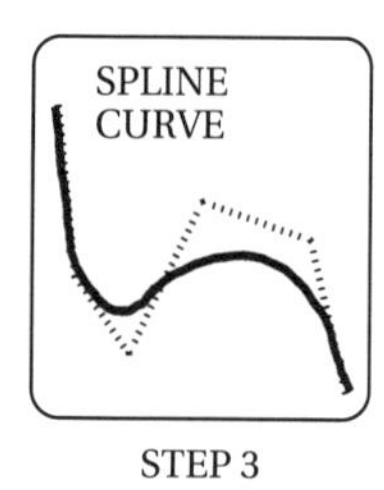

Figure 1.106 `PEDIT: Fit` and `Spline`.
Step 1 Draw a polyline (`PLINE`).
Step 2 `Command:` <u>`PEDIT`</u> (ENTER)
`Select polyline:` (Select line.)
`Close/Join/Width/Edit vertex/Fit/ Spline/Decurve/Ltype gen/Undo/eXit <X>:` <u>`Fit`</u>
Step 3 `PEDIT` with the `Spline` option for mathematical curve.

`Fit (F)` converts a polyline into a line composed of circular arcs that pass through each vertex (**Fig. 1.105**).

`Spline (S)` modifies the polyline as did the `Fit` curve, but it draws cubic curves passing through the first and last points, and not necessarily through the other points (**Fig. 1.106**).

`Decurve (D)` converts `Fit` or `Spline` curves to their original straight-line forms.

PEDIT: LINE GENERATE (MODIFY TOOLBAR)

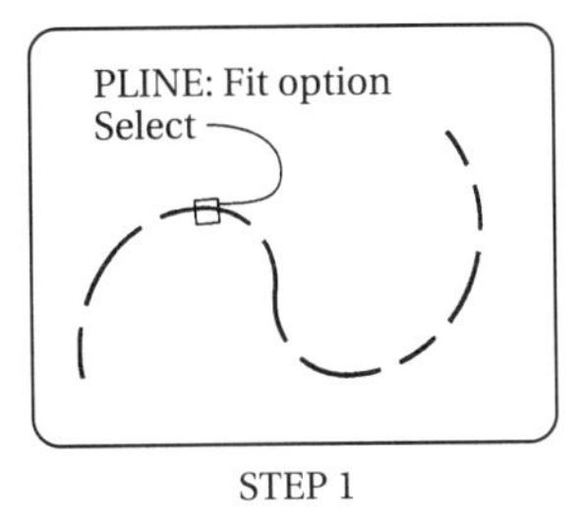

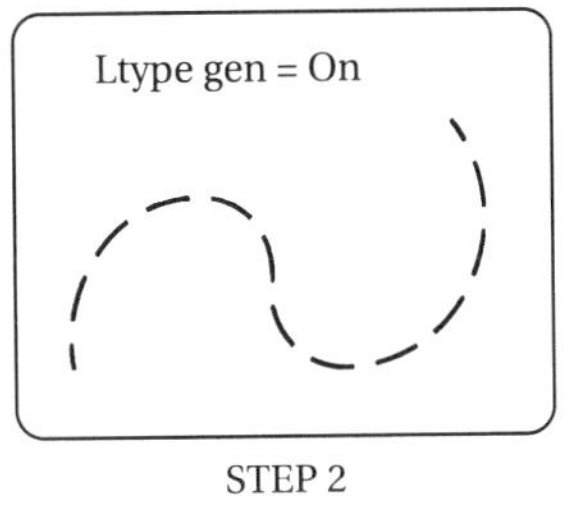

Figure 1.107 PEDIT: Ltype generate.
Step 1 Command: PEDIT (ENTER)
PEDIT Select polyline: (Select)
Close/Join/Width/Edit vertex/Fit/Spline/
Decurve/Ltype gen/Undo/eXit <X>: L (ENTER)
Step 2 Full PLINE line type ON/OFF <current>:
ON (ENTER) (Line is changed.)

PEDIT: ADD VERTEX

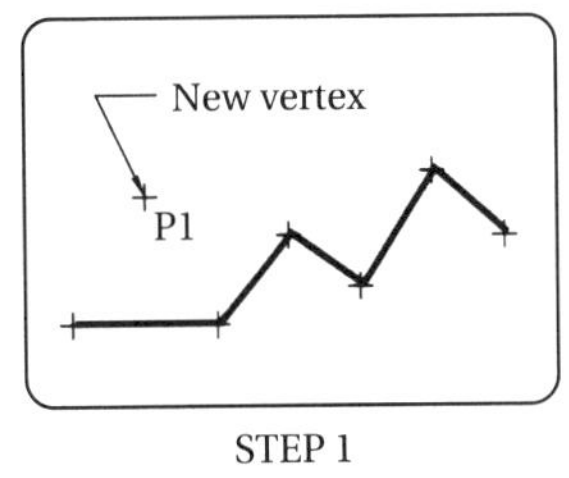

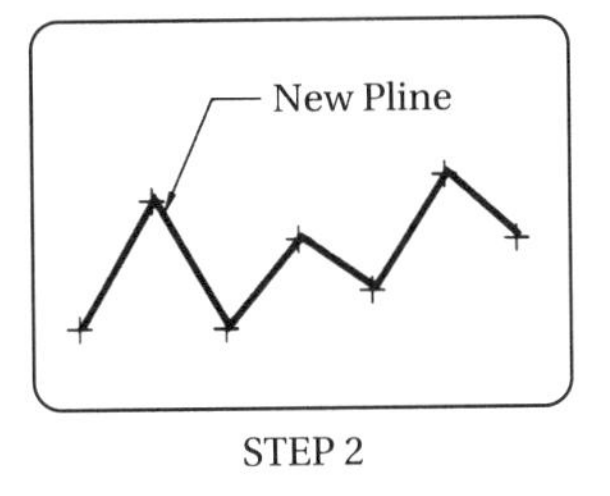

Figure 1.108 PEDIT: Edit Vertex, Insert.
Step 1 Use Edit Vertex option and place X on vertex before the new vertex.
Close/Join/Width/Edit vertex/Fit/Spline/
Decurve/Ltype gen/Undo/eXit <X>: Insert (ENTER)
Enter location of new vertex: P1
Step 2 Press (ENTER), the new vertex is inserted, and the polyline passes through it.

Ltype gen (L) applies dashed lines (such as centerlines) in a continuous pattern on curved polylines. Without applying this option, dashed lines may omit gaps in curved lines **(Fig. 1.107)**. By setting system variable Plinegen On, linetype generation will be applied as PLINES are drawn.

Undo (U) reverses the most recent PEDIT editing step. Edit Vertex (E) selects vertexes of the PLINE for editing by placing an X on the first vertex when the polyline is picked. The following options appear:

Next/Previous/Break/Insert/
Move/Regen/Straighten/Tangent/
Width/eXit/<N>: Next (ENTER)

PEDIT: MOVE VERTEX (MODIFY TOOLBAR)

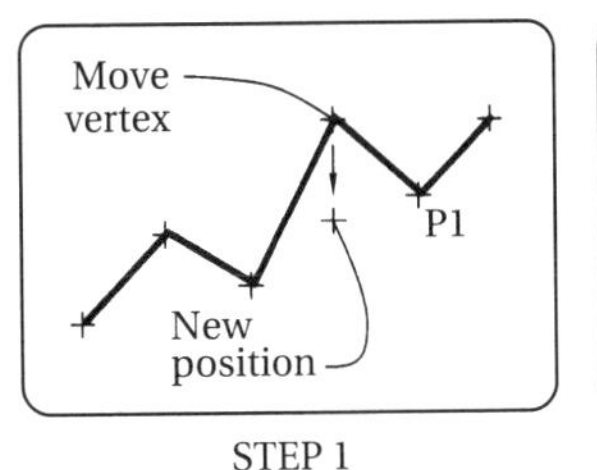

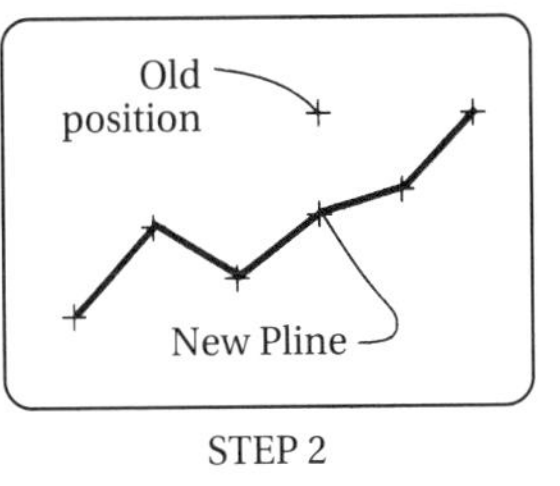

Figure 1.109 PEDIT: Edit Vertex, Move.
Step 1 Command: PEDIT (ENTER)
Select polyline: (Select line.)
Close/Join/Width/Edit vertex/Fit/
Spline/Decurve/Ltype gen/Undo/eXit <X>: E
((ENTER) X to vertex to be moved.)
Next/Previous/Break/Insert/Move/
Regen/Straighten/Tangent/Width/
eXit <N>: MOVE (ENTER)
Enter location of new vertex: P1
Step 2 Press (ENTER), and the polyline will be changed to pass through the moved vertex.

Next (N) and Previous (P) options move the X marker to next or previous vertices by pressing (ENTER). Break (B) prompts you to select a vertex with the X marker. Then use Next or Previous to move to a second point and pick Go to remove the line between the vertexes. Select eXit to leave the BREAK command and return to Edit Vertex.

Insert adds a new vertex to the polyline between a selected vertex and the next vertex **(Fig. 1.108)**. Move (M) relocates a selected vertex **(Fig. 1.109)**.

Straighten (S) converts the polyline into a straight line between two selected points. An X marker appears at the current vertex and the prompt, Next/Previous/Go/eXit/<N> appears. Move the X marker to a new vertex with Next or

PEDIT: STRAIGHTEN (MODIFY TOOLBAR)

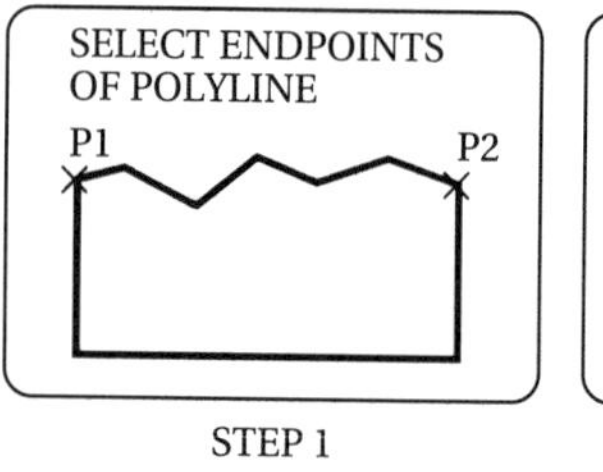

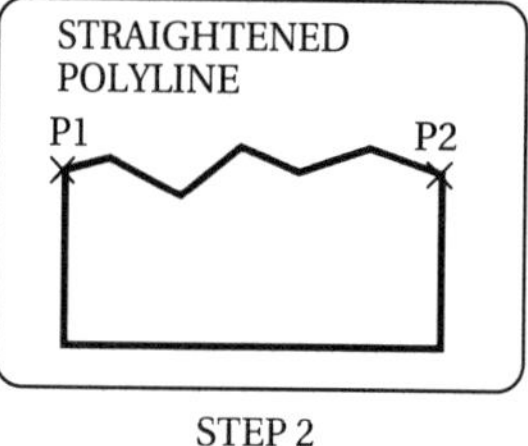

Figure 1.110 `PEDIT: Edit Vertex, Straighten.`
Step 1 Use `Edit Vertex` option of `PEDIT`, and place the `X` at the vertex where straightening is to begin, P1.
`Next/Previous/Break/Insert/Move/Regen/Straighten/Tangent/Width/eXit<N>:` Straighten (ENTER)
Step 2 `Next/Previous/Go/eXit <N>:` Next (ENTER) to P2
`Next/Previous/Go/eXit <N>:` Go (ENTER) (Line `P1-P2` is straightened.)

SPLINE COMMAND (DRAW TOOLBAR)

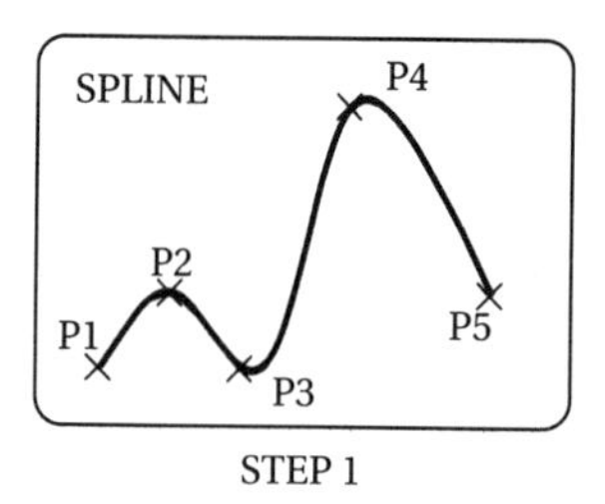

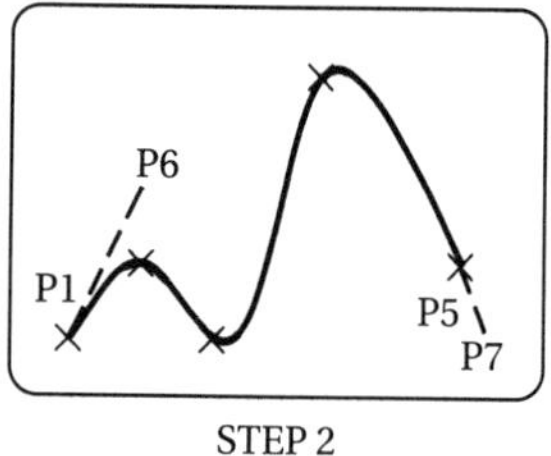

Figure 1.111 `SPLINE` command.
Step 1 `Command:` SPLINE (ENTER)
`Object/<Enter first point>:` P1
`Enter point:` P2
`Close/Fit Tolerance/<Enter point>:` P3, P4, P5 (ENTER)
`Close/Fit Tolerance/<Enter point>:` (ENTER)
Step 2 `Enter start tangent:` P6
`Enter end tangent:` P7

`Previous`, select `Go`, and the line is straightened between the vertices **(Fig. 1.110)**. Enter `X` to `eXit` and return to the `Edit Vertex` prompt.

`Tangent (T)` lets a tangent direction be selected at the vertex marked by the `X` for curve fitting by responding to the prompt `Direction of tangent`. Enter the angle from the keyboard or by cursor.

`Width (W)` sets the beginning and ending widths of an existing line segment from the `X`-marked vertex. Use `Next` and `Previous` to confirm in which direction the line will be drawn from the `X` marker. The polyline will be changed to its new thickness when the screen is regenerated with `Regen (R)`. Use `eXit` to escape from the `PEDIT` command.

1.34 SPLINE (Draw Toolbar)

`SPLINE` draws a smooth curve with a sequence of points within a specified tolerance as shown in **Fig. 1.111**. By setting `Fit Tolerance` to `0`, the curve will pass through the points; when set to a value greater than `0`, it will pass within a tolerance of each point. The `Endpoint Tangents` determine the angle of the spline at each end.

HATCH (DRAW TOOLBAR)

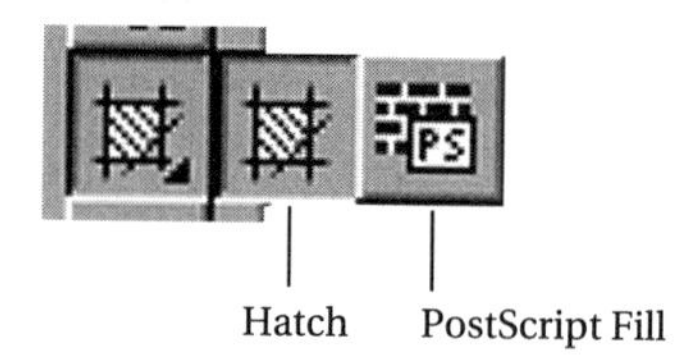

Figure 1.112 `HATCH` is one of the flyouts you can access from the `Draw` toolbar.

1.35 HATCHING (Draw Toolbar)

Hatching is a pattern of lines that fills sectioned areas, bars on graphs, and similar applications. From the `Draw` toolbar **(Fig. 1.112)** pick the Hatch icon and the `BHATCH` (boundary hatch) dialog box is displayed **(Fig. 1.113)**. By selecting the down arrow at the `Pattern` box, a drop-down list of pattern names is given from which to select. When one is selected, a view of the pattern will appear in the `Pattern-type` window. Examples of some predefined patterns are shown in **Fig. 1.114.**

The `Pattern Type` drop-down list box (Fig. 1.113) lets you specify `Predefined`, `User-defined`, or `Custom patterns`. Predefined patterns are those provided by AutoCAD.

BOUNDARY HATCH (BHATCH)

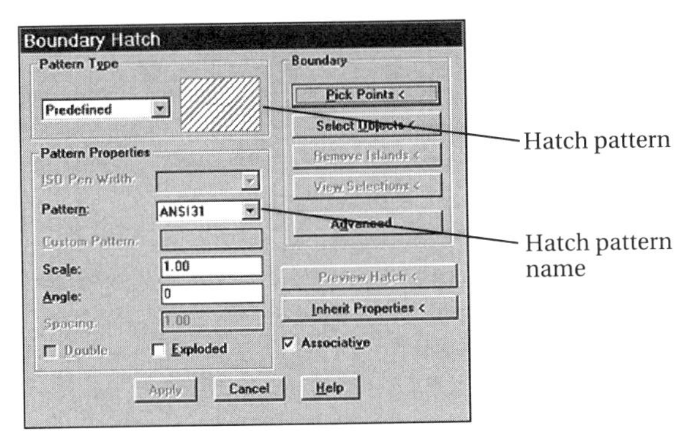

Figure 1.113 Select the `HATCH` icon or type `BHATCH` at the `Command` line to display this dialog box for setting hatching options.

HATCH PATTERN EXAMPLES

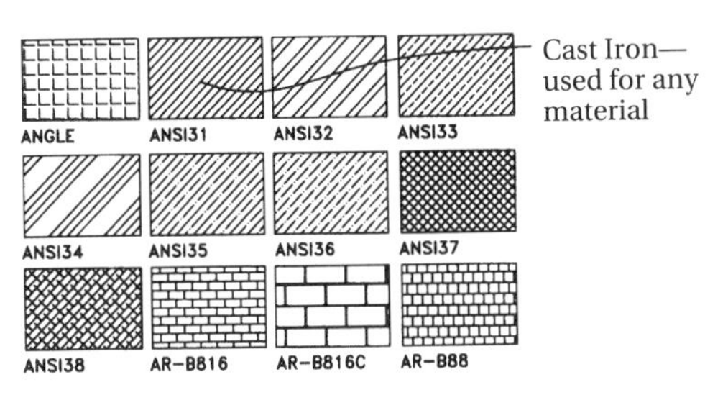

Figure 1.114 These are a few of the hatch patterns that are available.

`Scale` sets the spacing between the lines of a pattern, and `Angle` assigns their angle. The `Advanced` button displays the `Advanced Options` dialog box **(Fig. 1.115)** from which styles of `Normal`, `Outer`, or `Ignore` can be selected. A circle with a square and a triangle inside it illustrates the effect of each choice. `Normal` hatches every other nested area beginning with the outside. `Outer` hatches only the outside area, and `Ignore` hatches the entire area from the outer boundary ignoring any inner boundaries. When selected, `TEXT` within hatching appears in an opening in the hatching unless the `Ignore` style was selected.

ADVANCED HATCHING OPTIONS

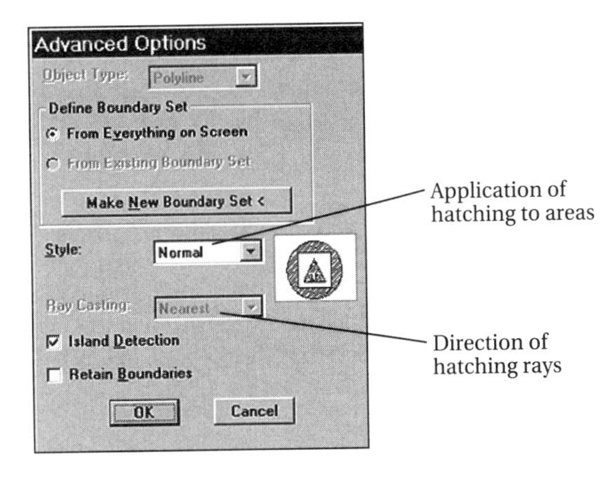

Figure 1.115 Select `Advanced` in the `Boundary Hatch` box (Fig. 1.113) to display this dialog box.

HATCHING AN AREA (DRAW TOOLBAR)

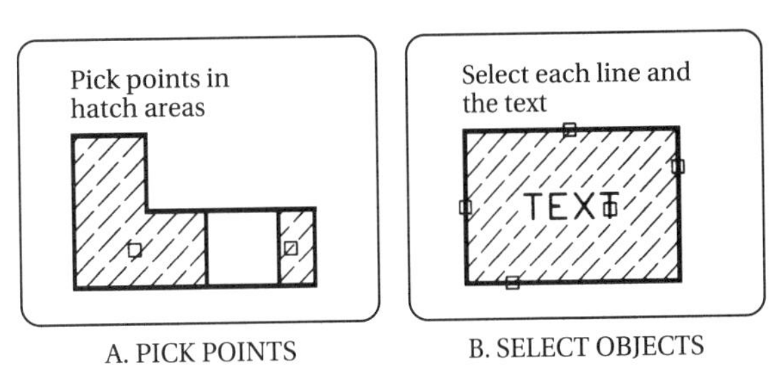

Figure 1.116 Hatching areas.
A The `Pick Points` option (Fig. 1.113) of `Boundary Hatch` prompts for points inside boundaries for hatching.
B The `Select Objects` option of hatching requires that the boundary lines be selected.

`Exploded Hatch` inserts a hatch pattern as a group of separate entities as if it were inserted as `*NAME` (`*ANSI31`, for example). `ISO Pen` enables you to set the width of a hatching line if a `Predefined ISO` pattern is selected (`ISO13W100`, for example). `Pick Points` and `Select Objects` (Fig. 1.113) are used to select areas inside of boundaries and then boundaries themselves **(Fig. 1.116)**. When points are selected outside the

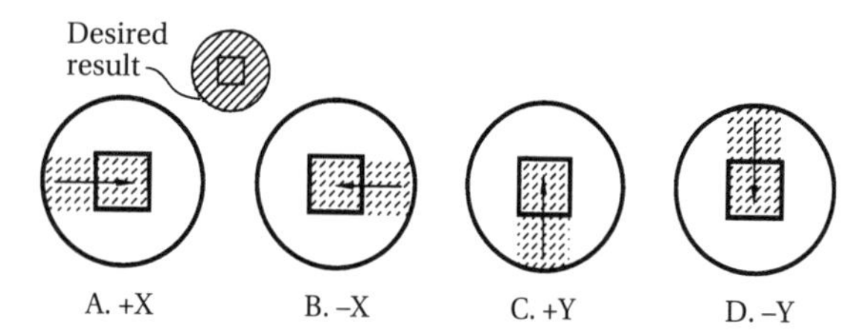

Figure 1.117 Select `Ray Casting` to control the direction a ray is cast to define a hatch boundary. When the directions above are specified, do not pick points in the shaded areas to avoid errors in boundary selection.

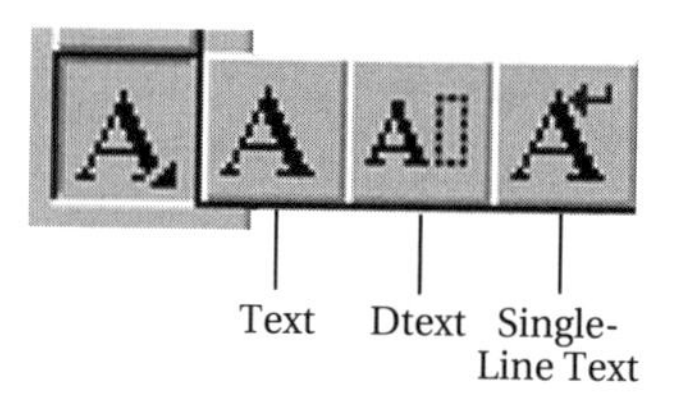

Figure 1.118 There are three `TEXT` flyouts you can access from the `Draw` toolbar.

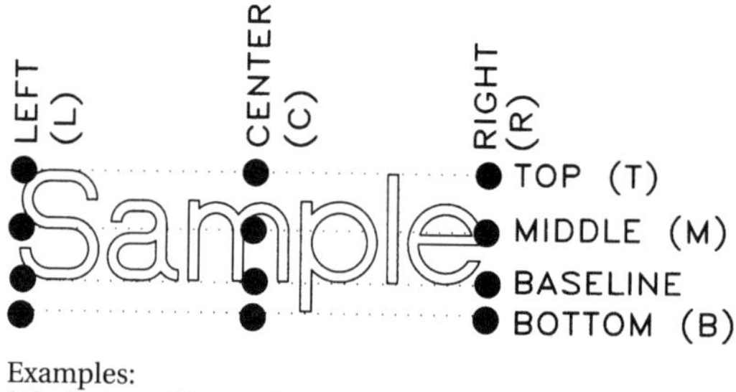

Figure 1.119 `Text` can be added to a drawing by using any of the insertion points above. For example, `BC` means the bottom center of a word or sentence that will be located at the cursor point.

boundary or if the boundary is not closed, error messages will appear.

`View Selections` shows all the boundaries and selected objects. `Preview Hatch` shows the result of the hatch selections before they are applied to allow changes in the hatch to be made and previewed. `Apply` draws hatching as a part of the drawing.

`Associative` is checked to associate hatching with its boundary so it will automatically change with any changes to the boundary.

The `Advanced Options` box (Fig. 1.115) gives additional options for hatching more efficiently. `From everything on the Screen` searches the whole screen during hatching.

`Make New Boundary Set` limits the search for areas to be hatched to a specified boundary set. `From Existing Boundary Set` is selected after a new boundary set is created and becomes the default until `BHATCH` is completed. `Retain Boundaries` keeps the defined boundary as a polyline in your drawing.

`Ray Casting` (available if `Island Detection` is disabled) specifies the direction rays are cast to seek a hatching boundary when `Pick Points` is used. `Nearest` casts a ray to the nearest hatchable object, turns left, and traces the boundary. To be sure that the correct boundary is selected, use the +X, –X, +Y, and –Y options to direct the ray's direction as it searches for the hatchable entity. **Figure 1.117** defines the areas in which pick points should not be selected for the desired results.

1.36 Text and Numerals (Draw Toolbar)

The `DTEXT` command, selected from the `Draw` toolbar **(Fig. 1.118)**, inserts text on the screen as it is typed:

```
Command: DTEXT (ENTER)
Justify/Style/<Start point>:
```

(Select point with cursor.)

```
Height <.18>: .125 (ENTER)
```

INSERTING DTEXT (DRAW TOOLBAR)

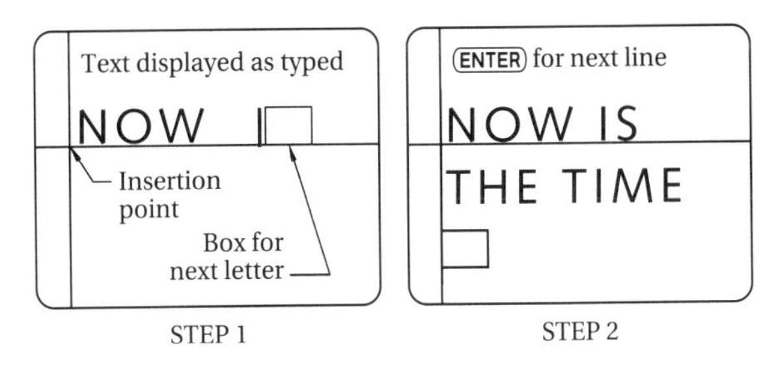

Figure 1.120 Inserting `DTEXT`.
Step 1 `Command:` DTEXT (ENTER)
`Justify/Style/<Start point>:` (Select point.)
`Height <.18>:` .125 (ENTER)
`Rotation angle <0>:` (ENTER)
`Text:` NOW IS
Step 2 Press (ENTER) and the box advances to next line of text, THE TIME (ENTER) (Box spaces down with each (ENTER).)

SPECIAL TEXT CHARACTERS

%%O	Start or stop Overline of text
%%U	Start or stop Underline of text
%%D	Degree symbol: 45%%D = 45°
%%P	Plus-minus: %%P0.05 = ±0.05
%%C	Diameter: %%C20 = ø20
%%%	Percent sign: 80%%% = 80%
%%nnn	Special character number nnn

Figure 1.121 These special characters beginning with `%%` are used while typing `DTEXT` and `TEXT` to obtain these symbols.

`Rotation angle <0>:` (ENTER)

`Text:` TYPE WORDS

`Justify` prompts for the insertion point for a string of text **(Fig. 1.119)**. `BC` means bottom center, `RT` means right top, and so forth.

Figure 1.120 illustrates how multiple lines of `DTEXT` are automatically spaced by pressing (ENTER) at the end of each line. The special characters shown in **Fig. 1.121** can be inserted by typing a double percent sign (`%%`) in front of them.

Type `QTEXT` and select `On` to reduce screen regeneration time by drawing text as boxes **(Fig. 1.122)**. When `QTEXT` is `Off`, the full text will be restored after regeneration (type `REGEN`).

QTEXT COMMAND

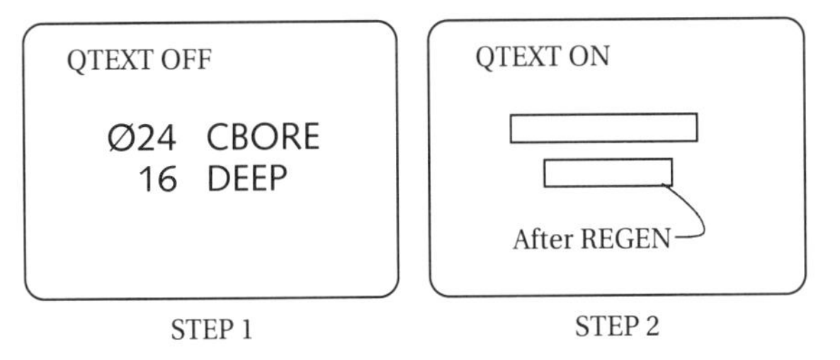

Figure 1.122 `QTEXT` command.
Step 1 `Command:` QTEXT (ENTER) `ON/OFF <CURRENT>:` ON (ENTER)
Step 2 `Command:` REGEN (ENTER) (Text is shown as boxes.)

1.37 Text STYLE (Data Menu)

Many of AutoCAD's text fonts and their names are shown in **Fig. 1.123.** The default style, `STANDARD`, uses the `TXT` font. New styles can be created as follows:

`Command:` STYLE (ENTER)

`Text style name (or ?) <STANDARD>:` PRETTY

`New style.`

The `Select Font File` dialog box is displayed **(Fig. 1.124)**. After you select one of AutoCAD's font files, the `STYLE` command continues:

`Height <0.20>:` 0 (ENTER)

`Width factor <1.0000>:` 1.25 (ENTER)

`Obliquing angle <0d>:` (ENTER)

`Backwards? <N>:` (ENTER)

`Upside-down? <N>:` (ENTER)

`Vertical? <N>:` (ENTER)

`PRETTY` is now the current text style.

The `Text Style` dialog box, displayed by selecting `Text Style...` from the `Data` pull-down

EXAMPLES OF FONTS

TXT	PRELIMINARY PLOTS
MONOTXT	FOR SPEED ONLY
	Simplex fonts
ROMANS	FOR WORKING DRAWINGS
SCRIPTS	Handwritten Style, 1234
GREEKS	ΓΡΕΕΚ ΣΙΜΠΛΕΞ, 12345
	Duplex fonts
ROMAND	THICK ROMAN TEXT
	Complex fonts
ROMANC	ROMAN WITH SERIFS
ITALICC	*ROMAN ITALICS TEXT*
SCRIPTC	Thick–Stroke Script Text
GREEKC	ΓΡΕΕΚ ΩΙΤΗ ΣΕΡΙΦΣ
	Triplex fonts
ROMANT	TRIPLE–STROKE ROMAN
ITALICT	*Triple–Stroke Italics*
	Gothic fonts
GOTHICE	English Gothic Text
GOTHICG	German Gothic Text, 12
GOTHICI	Italian Gothic Text, 12

Figure 1.123 Examples of available fonts are shown here.

menu, lists the defined text styles. An example of the text font is displayed when a style is selected **(Fig. 1.125)**. The style named PRETTY will retain its settings until they are changed. To select a different font for the PRETTY style, pick the Browse... button to the right of the Font File edit box to display the Select Font File dialog box (Fig. 1.124). Select a different font file and pick OK. Then pick the Apply button to apply the new font to any text already drawn using the PRETTY style. Pick Close to close the Text Style dialog box. This technique can be used to change text drawn using the TXT or MONOTXT font to a more attractive font at plot time. The TXT and MONOTXT fonts are often used initially in drawings because they regenerate more quickly.

SELECT FONT FILE (DATA MENU)

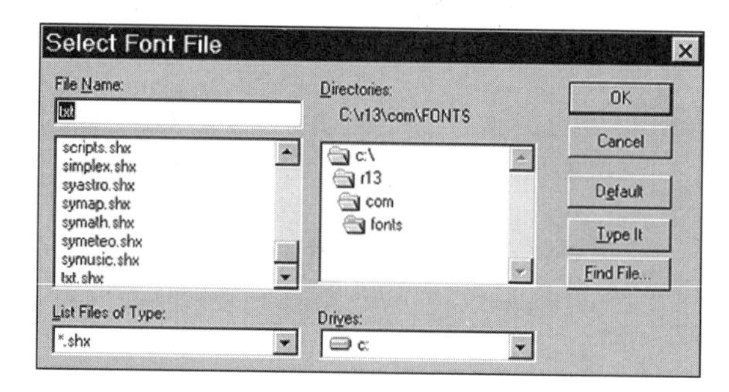

Figure 1.124 From the Data menu, select Text Style (or type STYLE at the Command line) to display this menu for selecting a font to be assigned to a text style.

OBJECT CREATION/TEXT STYLES

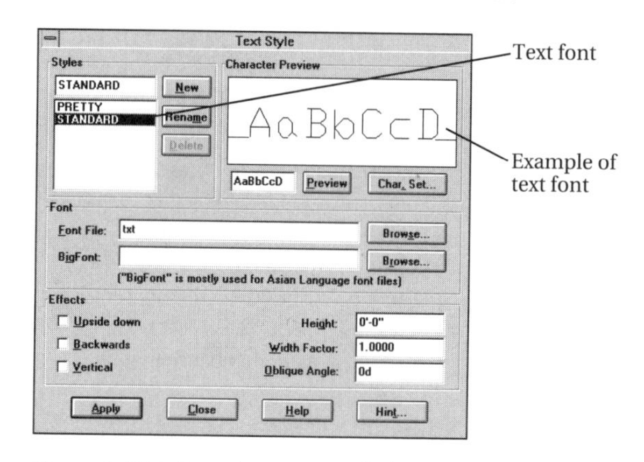

Figure 1.125 From the Data pull-down menu, select Text Style... to obtain this box. Select the name of the text style to be made current and information about the style will be displayed. Select Close to make it the current style.

DDEDIT BOX (MODIFY TOOLBAR)

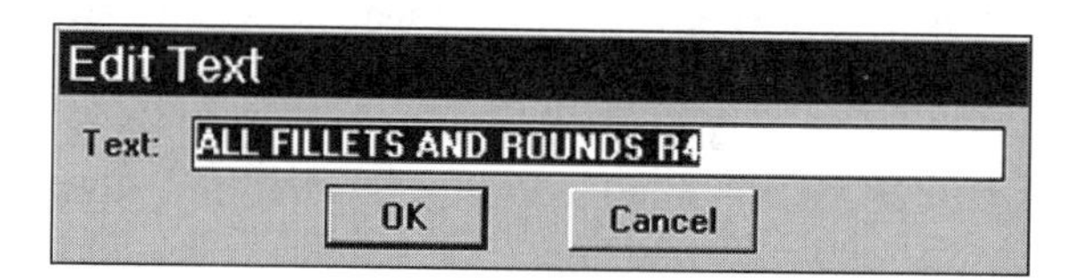

Figure 1.126 At the Command line, type DDEDIT and select a line of text on the screen. Text appears in the Edit Text box for modification.

EDIT MTEXT DIALOG BOX

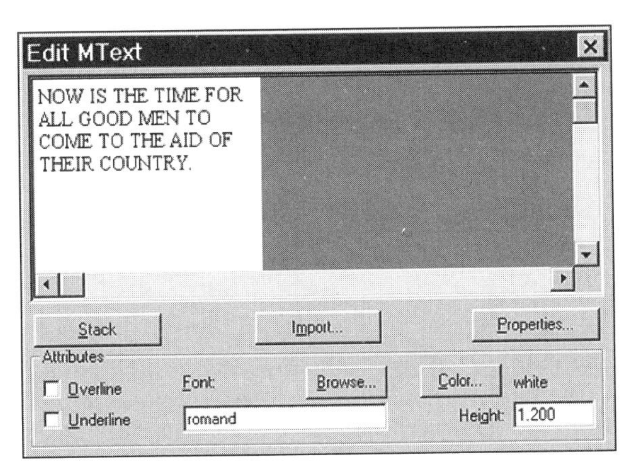

Figure 1.127 Type MTEXT at the Command line to obtain this dialog box and type multiple lines of text. If you use the DDEDIT command and select an existing MTEXT object, this dialog box is displayed. You can then make any changes needed.

MTEXT PROPERTIES

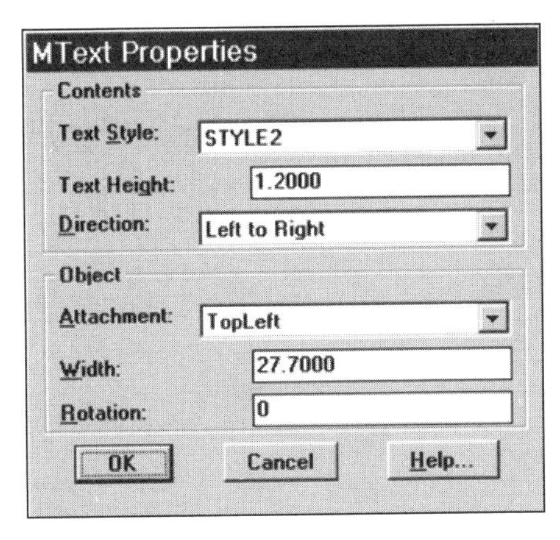

Figure 1.128 Select Properties from the Edit MText box to get this menu for editing MText properties.

EXAMPLES OF MTEXT

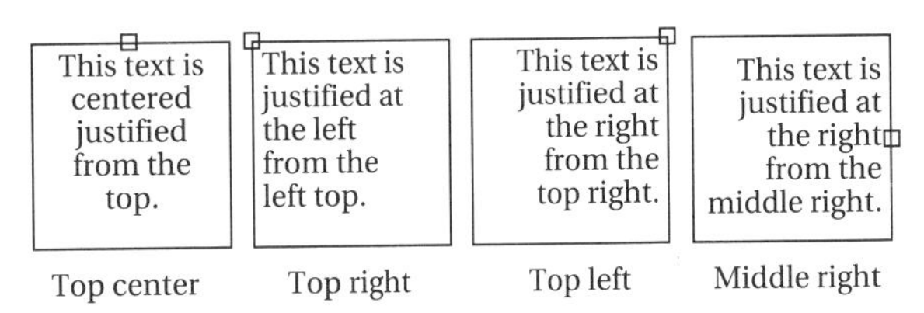

Figure 1.129 The Attach option of MText specifies the origin point of the text. TC is top center and TR is top right, for example.

The DDEDIT command is used to select a line of text to be displayed in a dialog box for editing **(Fig. 1.126)**. Correct the text, select the OK button, and it is revised on the screen.

1.38 MTEXT (Draw Toolbar)

From the Draw toolbar, select the Text icon, pick an insertion point, and specify the size of the text boundary by the diagonal window option, by specifying the width with W, or by specifying two points with 2P. The Edit Mtext dialog box will appear and the paragraph of text can be typed in its window **(Fig. 1.127)**. Select OK to place the paragraph on the drawing.

The following control keys can be used to edit the text:

CTRL + C	Copy text to Clipboard
CTRL + V	Paste Clipboard contents over selection
CTRL + X	Cut selection to the Clipboard
CTRL + Z	Undo and Redo
CTRL + SHIFT + SPACE	Insert nonbreaking space
ENTER	End current paragraph and start new line

Most of the options of the MText box are obvious. Stack lets you vertically align text as fractions would be drawn. Import displays a Text File dialog box for importing files from other sources.

Select Properties from the Mtext box to obtain the MText Properties dialog box **(Fig. 1.128)** for editing the text selected from the Edit MText box. Examples of text applied by the

MIRROR COMMAND

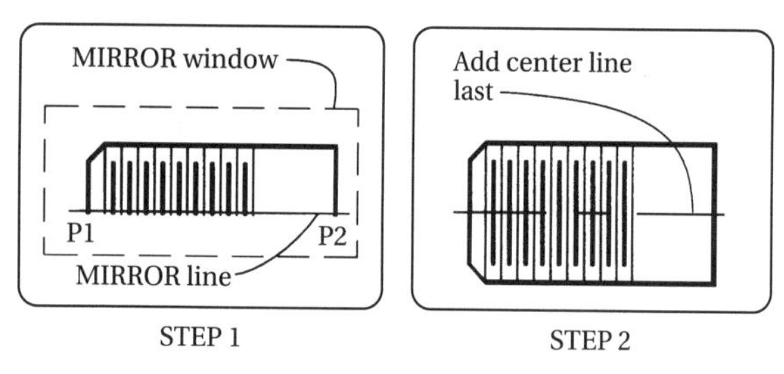

Figure 1.130 `MIRROR` command.
Step 1 `Command:` <u>`MIRROR`</u> (ENTER)
`Select objects:` <u>`W`</u> (ENTER) (Window drawing to be mirrored.)
`First point or mirror line: P1 Second point: P2`
Step 2 `Delete old objects? <N>:` <u>`N`</u> (ENTER) (The object is mirrored about the mirror line.)

OBJECT SNAP TOOLBAR

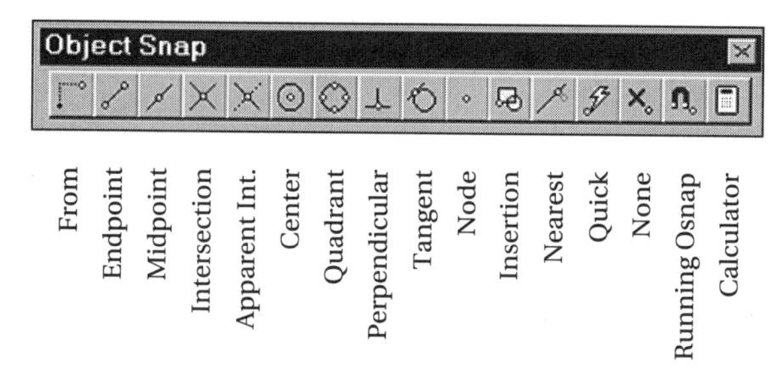

Figure 1.131 The `Object Snap` toolbar has these options for drawing to and from objects on the screen.

`Attach` option are shown in **Fig. 1.129**. `Attach` controls the text justification and spill relative to the text boundary.

1.39 MIRROR (Modify Toolbar)

Select the `MIRROR` icon from the `COPY` flyout on the `Modify` toolbar (or type `MIRROR` at the `Command` line) to `MIRROR` partial figures about an axis **(Fig. 1.130)**. A line that coincides with the `MIRROR` line (`P1-P2`, for example) will be drawn twice when mirrored; therefore, parting lines should be drawn after the drawing has been mirrored.

A system variable, `MIRRTEXT` under the `SETVAR` command, is used for mirroring text. By typing `SETVAR` and pressing (ENTER), `MIRRTEXT` (ENTER), and `0`, `MIRRTEXT` is set to `Off` and text will not be mirrored. If it is set to `1` (`On`), the text will be mirrored along with the drawing.

OSNAP: INTERSECTION & ENDPOINT

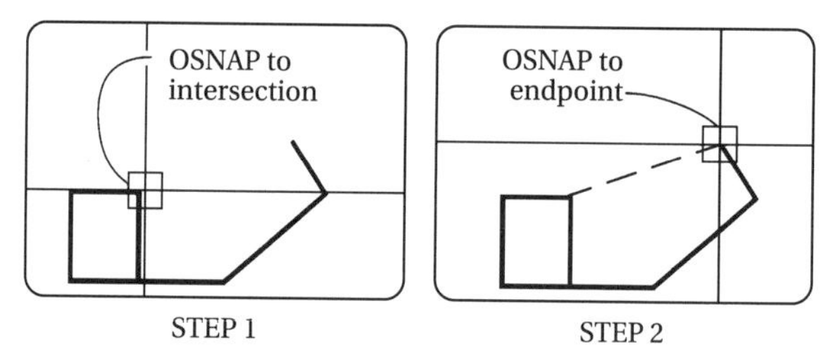

Figure 1.132 `OSNAP`: Intersection and end.
Step 1 `Command:` <u>`LINE`</u> (ENTER)
`From point:` (Select `Intersec` icon.)
`LINE From point:` <u>`int of`</u> (Select)
Step 2 `To point:` (Select `Endpoint` option of `OSNAP`.)
`To point:` <u>`endpoint of`</u> (Select endpoint and the line is drawn.)

1.40 OSNAP (Object Snap Toolbar)

By using `OSNAP` (`Object Snap`), you can snap to objects of a drawing rather than to the `grid`. `OSNAP` icons from the `Object snap` toolbar (**Fig. 1.131**) or the line of stars (`*****`) in the screen menu give the following options: `From`, `Endpoint`, `Midpoint`, `Intersection`, `Apparent Intersection`, `Center`, `Quadrant`, `Perpendicular`, `Tangent`, `Node`, `Insertion`, `Nearest`, `Quick`, and `None`. `OSNAP` is used as an accessory to other commands: `LINE`, `MOVE`, `BREAK`, and so forth.

Figure 1.132 shows how a line is drawn from an intersection to the endpoint of a line. In **Fig. 1.133**, a line is drawn from `P1` tangent to the circle by using the `Tangent` option of `OSNAP`. The `Tangent` option can be used to draw a line tangent to two arcs.

OSNAP: TANGENT OPTION (POINT TO CIRCLE)

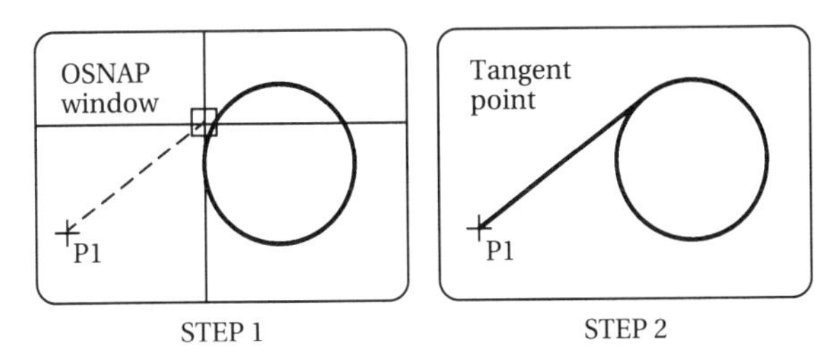

Figure 1.133 OSNAP: Tangent.
Step 1 `Command:` <u>LINE</u> (ENTER)
`From point:` <u>P1</u>
`To point:` (Select `Snap to tangent` icon.)
Step 2 `To point:` <u>tan to</u> (Select circle and the line is drawn tangent to it near the selected point.)

ARRAY: POLAR

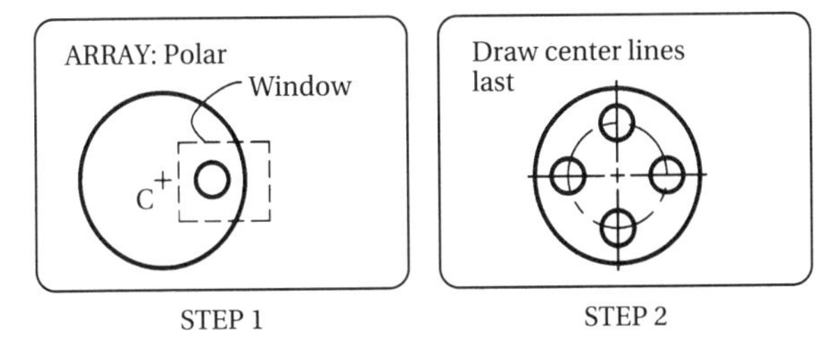

Figure 1.134 ARRAY: Polar.
Step 1 Drawing the figure to array.
`Command:` <u>ARRAY</u> (ENTER)
`Select objects:` <u>W</u> (Window the hole.) (ENTER)
`Select objects:` (ENTER)
`Rectangular or Polar array (R/P):` <u>P</u> (ENTER)
`Center point of array:` <u>C</u> (ENTER)
Step 2 `Number of items:` <u>4</u>
`Angle to fill (+=ccw, -=cw)<360>:` <u>360</u> (ENTER)
`Rotate objects as they are copied? <Y>` (ENTER) (The array is drawn.)

The `Node` option snaps to a `Point`, the `Quadrant` option snaps to one of the four compass points on a circle, the `Insert` option snaps to the intersection point of a `Block`, and the `None` option turns off `OSNAP` for the next selection. The `Quick` option reduces searching time by selecting

ARRAY: RECTANGULAR

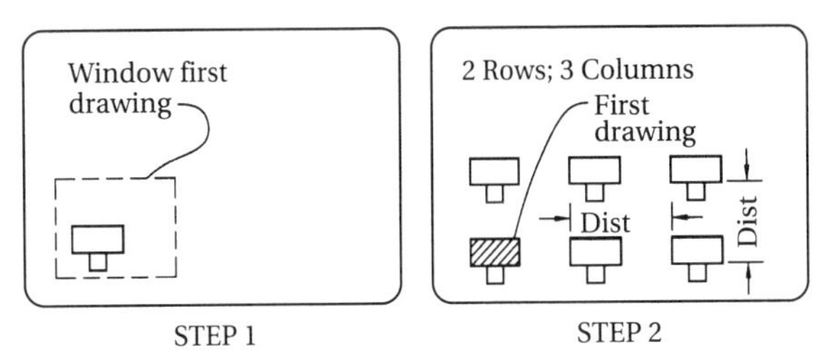

Figure 1.135 ARRAY: Rectangular.
Step 1 Draw desk in lower left of ARRAY.
`Command:` <u>ARRAY</u> (ENTER)
`Select Objects:` <u>W</u> (Window desk.) (ENTER)
`Rectangular or Circular array (R/P):` <u>R</u>
Step 2 `Number of rows (—) <1>:` <u>2</u> (ENTER)
`Number of columns (<vb><vb><vb>) <1>:` <u>3</u> (ENTER)
`Unit cell distance between rows (—):` <u>4</u> (ENTER)
`Distance between columns (<vb><vb><vb>):` <u>3.5</u> (ENTER) (ENTER) (The array is drawn.)

the first object encountered rather than searching for the one closest to the aperture's center.

`OSNAP` settings can be temporarily retained as "running" `OSNAP`s for repetitive use. To set running `OSNAP`s to `Endpoints` and circle `Centers`, do the following:

`Command:` <u>OSNAP</u> (ENTER)

`Object snap modes:` <u>ENDPOINT, CENTER</u> (ENTER)

Now, the cursor has an aperture target at its intersection for picking endpoints and centers of arcs. Remove running `OSNAP` settings by typing `OSNAP` and pressing (ENTER). Running `OSNAPS` can be set by selecting the icon on the `Object Snap` toolbar. The `APERTURE` command sets the size to the target box that appears at the cursor for `OSNAP` applications. Its size can vary from 1 to 50 pixels square.

1.41 ARRAY (Modify Toolbar)

Select `ARRAY` (`Polar Array` or `Rectangular Array`) from the `COPY` flyout of the `Modify` toolbar to draw circular or rectangular patterns (rows and columns) of selected drawings, respectively.

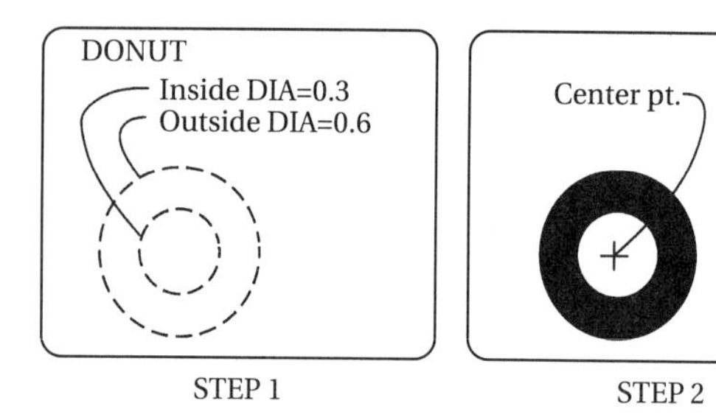

Figure 1.136 DONUT command.
Step 1 Command: DONUT (ENTER)
Inside diameter <0.0000>:.30 (ENTER)
Outside diameter <0.0000>:.60 (ENTER)
Step 2 Center of donut: (Select point, and donut is drawn.)

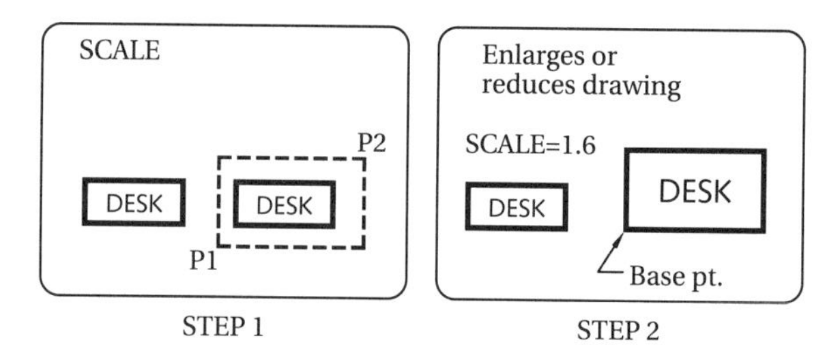

Figure 1.137 SCALE command.
Step 1 Command: SCALE (ENTER)
Select objects: (Window with P1 and P2)
Step 2 Base point: (Select base point.)
<Scale factor>/Reference: 1.6 (ENTER) (The desk is drawn 60% larger.)

A series of holes can be located on a bolt circle by drawing the first hole and ARRAYing it as a **polar array (Fig. 1.134)**.

A **rectangular array** is begun by making the drawing in the lower left corner and following the steps in **Fig. 1.135**. Rectangular arrays may be drawn at angles by using the SNAP command to ROTATE the grid. The first object is drawn in the lower left corner of the array and the number of **rows**, **columns**, and the **cell distances** are specified when prompted.

1.42 DONUT (Draw Toolbar)

The DONUT command, selected from the Circle flyout of the Draw toolbar, draws doughnuts by assigning outside and inside diameters **(Fig. 1.136)**. DONUTs can be drawn as solid circles by setting the inside diameter to 0.

1.43 SCALE (Modify Toolbar)

The SCALE command reduces or enlarges previously drawn objects. The desk in **Fig. 1.137** is enlarged by windowing it, selecting a base point, and typing a scale factor of 1.6. The drawing and its text are enlarged in the X- and Y-directions.

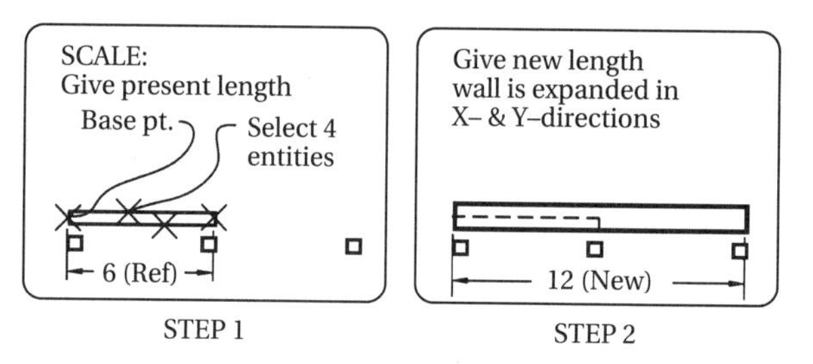

Figure 1.138 SCALE: Reference dimensions.
Step 1 Command: SCALE (ENTER)
Select objects: (Select points on each line.)
Base point: (Select point.)
<Scale factor>/Reference: R (ENTER)
Reference length <1>: 6 (ENTER)
Step 2 New length: 12 (ENTER) (The drawing is enlarged in all directions.)

A second option of SCALE lets you select a length of a given figure, specify its present length, and assign a length as a ratio of the first dimension **(Fig. 1.138)**. The lengths can be given by the cursor or typed at the keyboard in numeric values.

STRETCH COMMAND

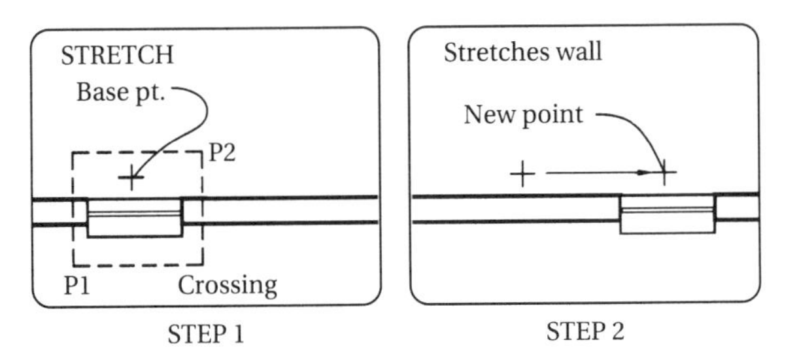

Figure 1.139 STRETCH command.
Step 1 Command: STRETCH
Select objects to stretch by crossing-window or -polygon...
Select objects: C (crossing window) ENTER
First corner: P1
Other corner: P2
Select objects: ENTER
Base point: (Select base point.)
Step 2 New point: (Select new point.)
(The symbol is repositioned.)

1.44 STRETCH (Modify Toolbar)

The STRETCH command lengthens or shortens a portion of a drawing while one end is left stationary. The window symbol in the floor plan **Fig. 1.139** is STRETCHed to a new position, leaving the lines of the wall unchanged. A Crossing window must be used to select lines that will be stretched.

1.45 ROTATE (Modify Toolbar)

An object can be rotated about a base point by using the ROTATE command **(Fig. 1.140)**. Window the object, select a base point, and type the rotation angle or select it with the cursor. Drawings made on multiple layers can be rotated also.

1.46 SETVAR (Command Line)

Many **system variables** can be inspected by typing SETVAR and ? at the Command line, and changed if they are not read-only variables. To change one or more variables (TEXTSIZE, for example), respond as follows:

Command: SETVAR ENTER

Variable name or ?: TEXTSIZE ENTER

New value for TEXTSIZE <0.18>: 0.125 ENTER

By entering the SETVAR command with an apostrophe in front of it ('SETVAR), it can be used transparently without leaving the command in progress.

ROTATE COMMAND

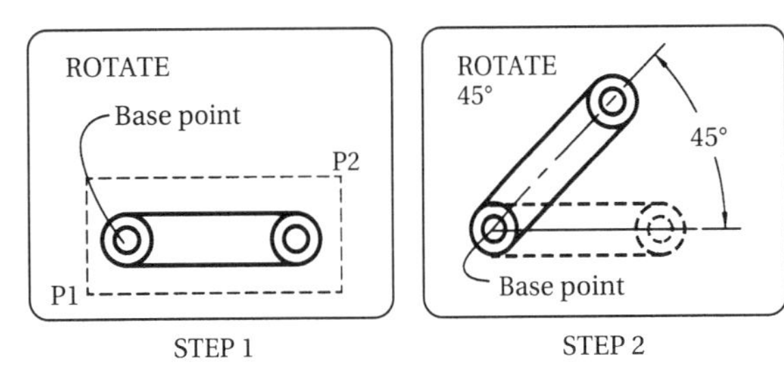

Figure 1.140 ROTATE command.
Step 1 Command: ROTATE ENTER
Select objects: W ENTER P1 and P2
Step 2 Base point: P3
<Rotation angle>/Reference: 45 ENTER
(Object is rotated 45° CCW.)

DIVIDE COMMAND

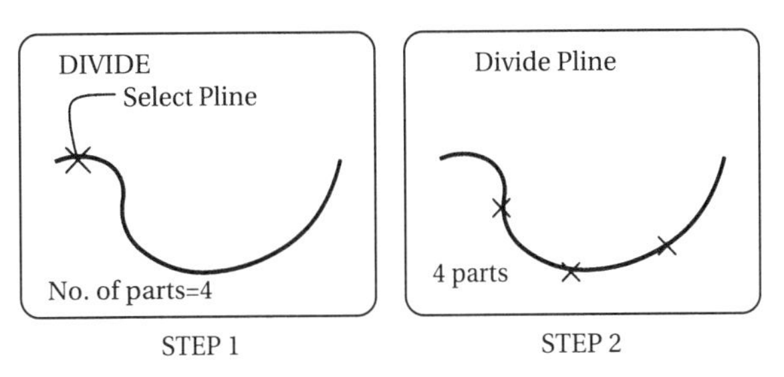

Figure 1.141 DIVIDE command: line.
Step 1 Command: DIVIDE ENTER
Select object to divide: (Select Pline.)
Step 2 <Number of segments>/Block: 4 ENTER
(PDMODE symbols are placed along the line, dividing it.)

DIVIDE COMMAND

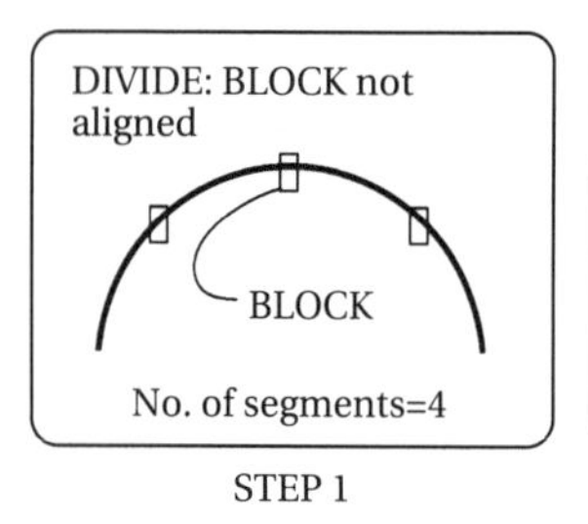

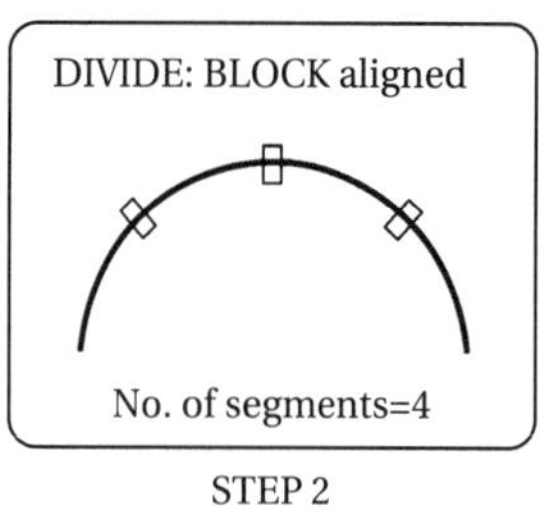

Figure 1.142 `DIVIDE`: arc.
Step 1 `Command:` DIVIDE ENTER
`Select object to divide:` (Select arc.)
`<Number of segments>/Block:` B ENTER
`Block name to insert:` RECT ENTER
`Align block with object? <Y>` N ENTER
`Number of segments:` 4 ENTER
Step 2 `Align block with object? <Y>` ENTER
(The blocks radiate from the arc's center.)

1.47 DIVIDE (Draw Toolbar)

The `DIVIDE` command places markers on a line to show a specified number of equal divisions. The polyline in **Fig. 1.141** is selected by the cursor, the number of divisions is specified, and markers are equally spaced along it. The markers will be of the type and size currently set by the `PDMODE` and `PDSIZE` variables under the `SETVAR` command.

The `Block` option of `DIVIDE` allows saved blocks (rectangles in this example) to be used as markers on the line **(Fig. 1.142).** Blocks can be either `Aligned` or `Not Aligned` as shown.

1.48 MEASURE (Draw Toolbar)

The `MEASURE` command repeatedly measures off a specified distance along an arc, circle, polyline, or line and places markers at these distances **(Fig. 1.143)**. Respond to the `Select object to measure` prompt by picking a point near the end where measuring is to begin. When prompted, give the segment length, and markers are displayed along the line. The last segment is usually a shorter length.

MEASURE COMMAND

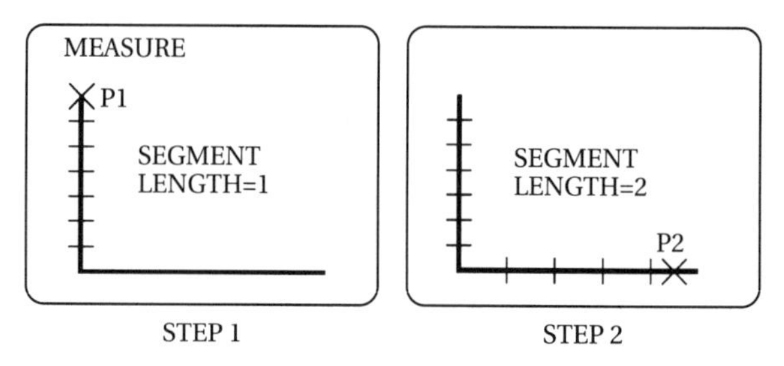

Figure 1.143 `MEASURE` command.
Step 1 `Command:` MEASURE ENTER
`Select object to measure:` P1
`<Segment length>/Block:` 0.1 ENTER
(The line is divided into 0.1 divisions starting at the end nearest `P1`.)
Step 2 `Command:` MEASURE ENTER
`Select object to measure:` P1
`<Segment length>/Block:` 0.2 ENTER (0.2 divisions are located along the line.)

OFFSET COMMAND

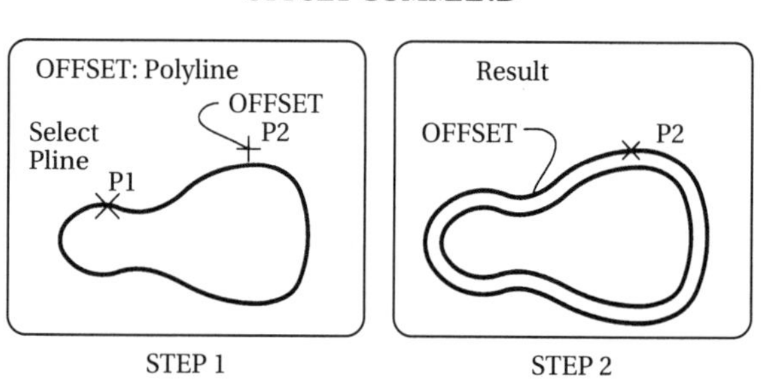

Figure 1.144 `OFFSET` command.
Step 1 `Command:` OFFSET ENTER
`Offset distance or Through <Through>:` T ENTER
`Select object to offset:` P1
`Through point:` P2
Step 2 An enlarged `Pline` is drawn that passes through `P2`.

1.49 OFFSET (Modify Toolbar)

An object can be drawn parallel to and offset from another object, such as a polyline, by the `OFFSET` command **(Fig. 1.144).** `OFFSET` prompts for the distance or the point through which the offset line must pass, and then prompts for the side of the off-

BLOCK COMMAND

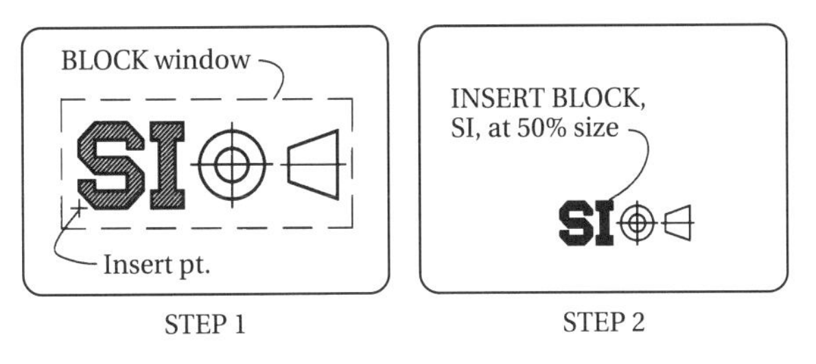

Figure 1.145 BLOCK command.
Step 1 Command: BLOCK (ENTER)
Make a drawing to be BLOCKed, and respond to the BLOCK command as follows:
BLOCK name (or ?): SI (ENTER)
Insertion base point: (Select insert point.)
Select objects: WINDOW or W (ENTER)
(Window the drawing, and it disappears into memory.)
Step 2 Command: INSERT (ENTER)
Block name (or ?): SI (ENTER)
Insertion point: (Select point with cursor.)
X-scale factor <1>/Corner/XYZ: 0.5 (ENTER)
Y-scale factor <default=X>: (ENTER)
Rotation angle <0>: (ENTER) (Block SI is inserted at 50% size.)

INSERT (DRAW TOOLBAR)

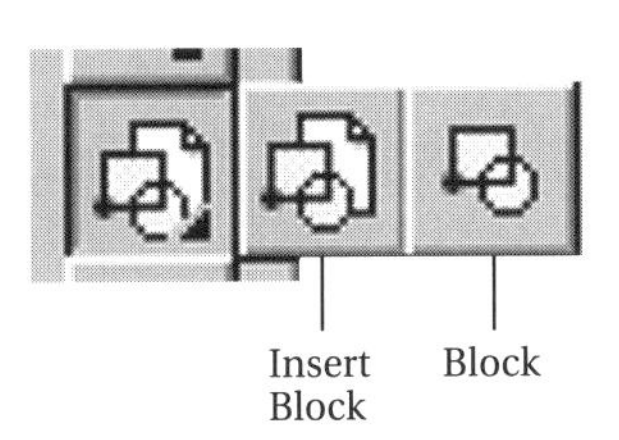

Figure 1.146 From the Draw toolbar, two BLOCK options may be selected.

INSERT DIALOG BOX (DDINSERT)

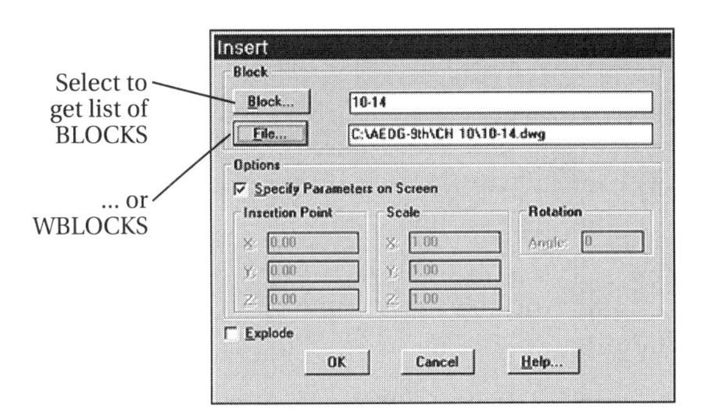

Figure 1.147 Select the BLOCK flyout from the Draw toolbar to specify BLOCKs to be INSERTed.

SELECT A DRAWING FILE BOX

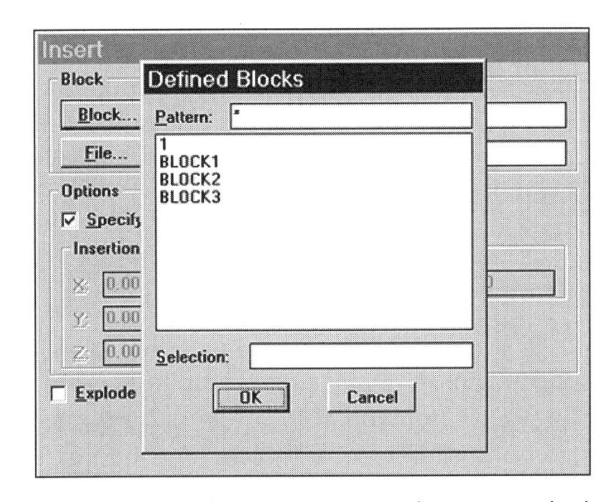

Figure 1.148 When INSERT is used, some standard Defined BLOCKS are provided from which to select.

set. The OFFSET command is helpful when drawing parallel lines to represent walls of a floor plan.

1.50 BLOCKs (Draw Toolbar)

One of the more productive features of computer graphics is the capability of creating objects called BLOCKs for repetitive use. Objects, such as the SI symbol, are drawn, made into BLOCKs, and inserted into drawings as shown in **Fig. 1.145**. The INSERT command, on the Draw toolbar **(Fig. 1.146)**, gives the Insert dialog box **(Fig. 1.147)** for inserting both BLOCKs and Files (WBLOCKs). Select BLOCK and the Select a Drawing File dialog box appears for selection of a BLOCK **(Fig. 1.148)**.

When a BLOCK is selected, it is selected as a total unit. However, BLOCKs that were INSERTed by selecting the Explode check box first, or by typing a star in front of the BLOCK name (*SI, for example), can be selected one object at a time. An inserted BLOCK can be separated into individual entities by typing EXPLODE and selecting the BLOCK.

BLOCKs can be used only in the current drawing file unless they are converted to WBLOCKs (Write BLOCKs), which become independent files, not parts of files. This conversion is performed as follows:

Command: WBLOCK (ENTER)

File Name: B:SI (ENTER) (This assigns the name of the WBLOCK to drive A.)

Block Name: SI (ENTER) (This is the name of the BLOCK that is being changed to a WBLOCK.)

BLOCKs can be redefined by selecting a previously used BLOCK name to receive the prompt, Redefine it? <N>:. Type Y (Yes) and select the new objects to be blocked. After doing so, the redefined BLOCKs automatically replace those in the current drawing with the same name.

1.51 Transparent Commands (Command Line)

Transparent commands are commands used while another command is in progress by typing an apostrophe in front of the command name at the Command line. If you are dimensioning a part and wish to use PAN, type 'PAN, press (ENTER), do the pan, and then complete the dimensioning command. Commands that can be used transparently are 'ABOUT, 'APERTURE, 'APPLOAD, 'ATTDISP, 'BASE, 'BLIPMODE, 'CAL, 'COLOR, 'DDEMODES, 'DDGRIPS, 'DDRMODES, 'DDSELECT, 'DDSTYLE, 'DELAY, 'DIST, 'DRAGMODE, 'ELEV, 'FILES, 'FILL, 'GIFIN, 'GRAPHSCR, 'GRID, 'HELP, 'ID, 'ISOPLANE, 'LAYER, 'LIMITS, 'LINETYPE, 'LTSCALE, 'MTPROP, 'ORTHO, 'PAN, 'PCXIN, 'QTEXT, 'REDRAW, 'REDRAWALL, 'REGENAUTO, 'RESUME, 'SCRIPT, 'SETVAR, 'SPELL, 'XSTATUS, 'STYLE, 'TEXTSCR, 'TIFFIN, 'TIME, 'TREESTAT, 'UNITS, 'VIEW, and 'ZOOM.

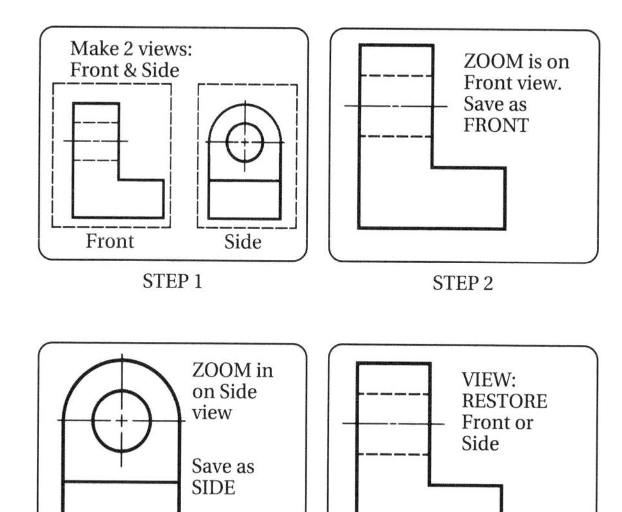

Figure 1.149 VIEW command.
Step 1 The two-view drawing can be saved as separate views.
Step 2 ZOOM the front view to fill the screen.
Command: VIEW (ENTER)
?/Delete/Restore/Save/Window: S (ENTER)
View name to save: FRONT (ENTER)
Step 3 ZOOM the side view to fill the screen.
Command: VIEW (ENTER)
?/Delete/Restore/Save/Window: S (ENTER)
View name to save: SIDE (ENTER)
Step 4 To display a view: Command: VIEW (ENTER)
?/Delete/Restore/Save/Window: R (ENTER)
View name to save: FRONT (ENTER) (View is displayed on screen.)

1.52 VIEW (Command Line)

Portions of drawings can be saved as separate views with the VIEW command **(Fig. 1.149)**. The entire screen can be made into a VIEW by the Save option and naming it when prompted. The Window option makes a VIEW of the windowed portion of the drawing. Type RESTORE and give the VIEW's name and press (ENTER) to display it. Type ? to list the saved VIEWS and select DELETE to remove a VIEW from the list.

LIST COMMAND

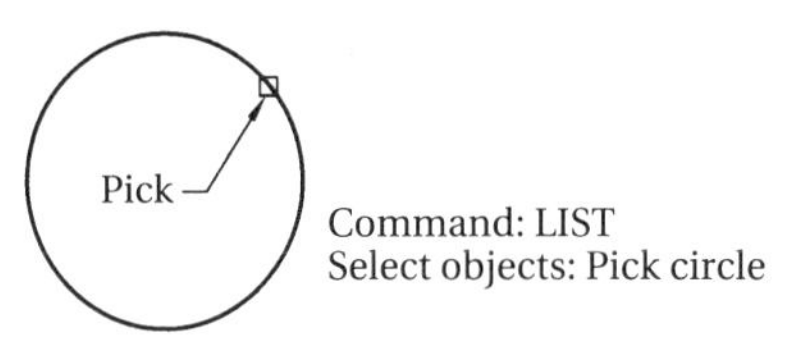

CIRCLE Layer:	VISIBLE
Space:	Model space
Center point:	X = 2.00 Y = 1.00 Z = 0.00
Radius:	1.00
Circumference:	6.28
Area:	3.14

Figure 1.150 Type `LIST` at the `Command` line to obtain information about objects on the screen, such as the circle shown here.

1.53 Inquiry Commands (Object Properties Toolbar)

From the inquiry command flyout on the `Object Properties` toolbar, you can select the `List`, `ID`, `Distance`, `Area`, or `Massprop` commands to obtain information about objects in the drawing. You can also type these commands as well as other inquiry commands such as `Dblist`, `Status`, and `Time` to obtain information about the current drawing file. Selecting the `List` command and then picking a circle (or any other object) displays information about the selected object **(Fig. 1.150)**. The `Dblist` command displays a listing of every object in the current drawing. If the list is very long, it will pause between screens. Pressing (ESC) aborts the command.

`Dist` measures the distance, its angle, and its delta-X and delta-Y distances between selected points without drawing a line. `ID` gives the X-, Y-, and Z-coordinates of a point that is picked on the screen. `Area` gives the perimeter and area space on the screen. Prompts request `First point:`, `Next point:`, `Next point:`, and so on to pick all points; then press (ENTER). `Areas` can be added and removed when they are being selected as shown in **Fig. 1.151**.

The `Status` option gives information about the settings, layers, coordinates, and disk space.

CALCULATION OF AREAS

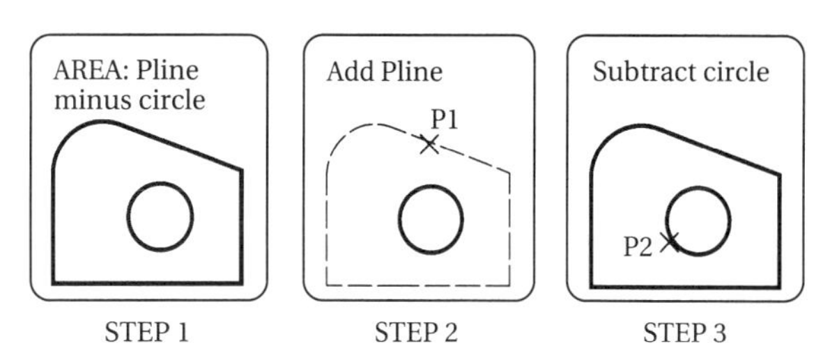

Figure 1.151 `AREA` command.
Step 1 The object drawn as a `Pline` with a circular area can have its area determined with the `AREA` command.
Step 2 `Command: AREA`
`<First point>/Object/Add/Subtract: A`
`<First point>/Object/Subtract: O`
`(ADD mode) Select object: P1`
(The pline area is displayed.)
Step 3 `(ADD mode) Select objects:` (ENTER)
`<First point>/Object/Subtract: S`
`<First point>/Object/Add: O`
`(SUBTRACT mode) Select objects: P2`
(The circle's area and the total area are displayed.)

TIME COMMAND (DATA MENU)

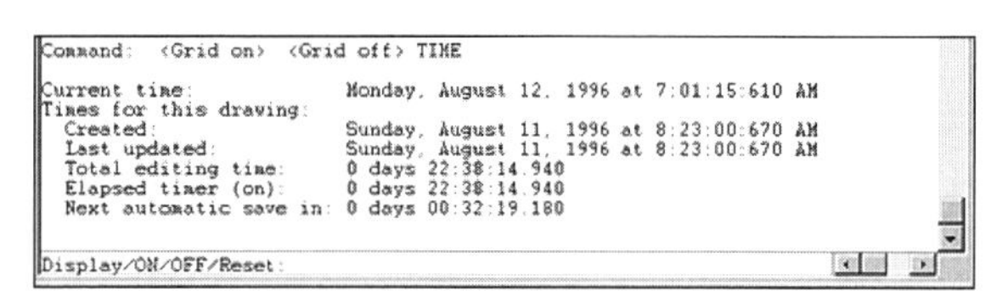

Figure 1.152 The `Time` option under the `Data` menu can be used for inspecting the time spent on a drawing and for setting the time for an assignment.

`Time` displays information about the time spent on a drawing **(Fig. 1.152)**. The timer can be `RESET` and turned `On` to record the time of a drawing session, but the cumulative time cannot be erased without deleting the drawing file. After `RESET`ting, `Display` shows the time of the current session opposite the heading, `Elapsed time:`. By selecting `Utilities...` from the `Management` flyout menu in the `File` pull-down menu, the `File Utility` dialog box is displayed **(Fig. 1.153)**.

FILE UTILITIES

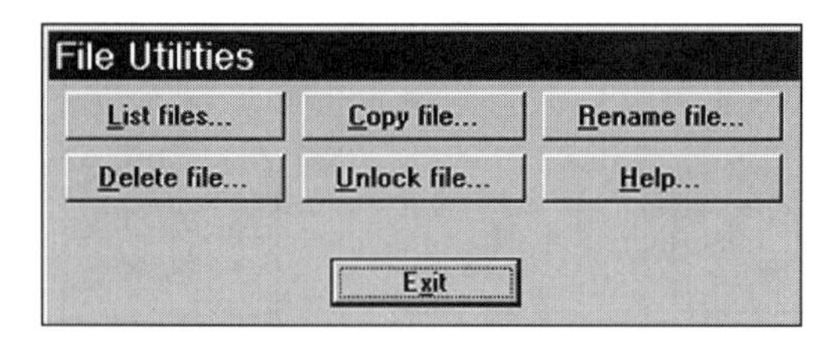

Figure 1.153 Under File, Management, and Utilities, this menu lets you make changes in files.

TYPES OF DIMENSIONS

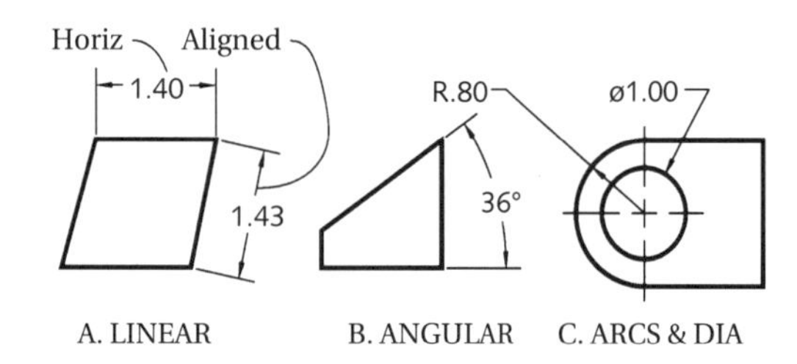

Figure 1.154 The types of dimensions that appear on a drawing.

1.54 Dimensioning

Figure 1.154 shows common types of dimensions that are applied to drawings. Drawings should be drawn full size on the screen since they are measured by the dimensioning commands; scaling should be done at plotting time.

Dimensions can be applied as **associative** or as **nonassociative** (**exploded**) dimensions. Associative dimensions (when DIMASO is On) are inserted as if the dimension line, extension lines, text, and arrows were parts of a single BLOCK. Exploded dimensions are applied as individual objects that can be modified independently by setting DIMASO to Off. Except where noted, the examples that follow will be associative dimensions.

DIMSTYLE (COMMAND LINE)

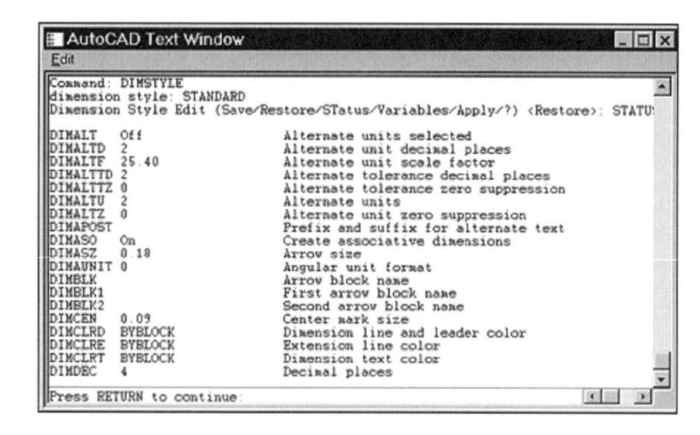

Figure 1.155 Type DIMSTYLE at the Command line, STATUS, and press (ENTER) to obtain a listing of dimension system variables.

BASIC DIMENSIONING VARIABLES

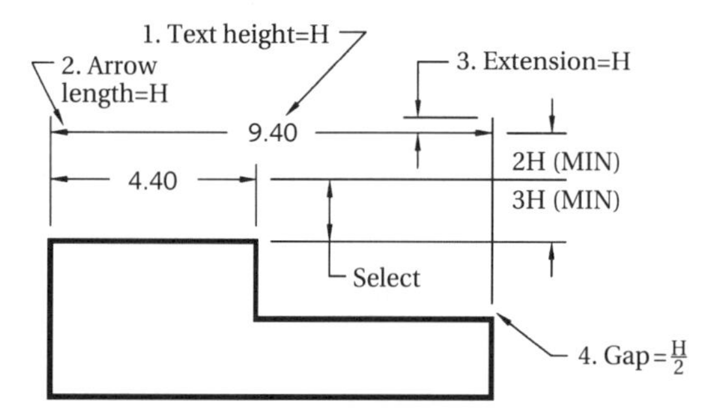

DIM VARS
1 DIMTXT (TEXT HT.)=H=.125
2 DIMASZ (ARROW)=H=.125
3 DIMEXE (EXTENSION)=H=.125
4 DIMEXO (OFFSET)=H=H/2=.06
5 DIMDLI (INCREMENT)=2H=.25 MIN.
6 DIMSCALE (SCALE)=1 for INCHES
25.4 for MILLIMETERS

Figure 1.156 Dimensioning variables are based on the height of the lettering (text), usually about $\frac{1}{8}$ inch.

Many variables must be set before dimensioning is usable: Arrowheads and numerals must be sized, extension line offsets specified, text fonts assigned, and units adopted, to name a few.

DIMENSIONING SYSTEM VARIABLES

DIM VARS	DEFAULT	DEFINITION
DIMALT	OFF	Alternate units selected
DIMALTD	2	Alternate units decimal pls.
DIMALTF	25.4	Alternate units scale factor
DIMALTTD	2	Alternate tolerance dec. pl.
DIMALTTZ	0	Alternate toler. zero suppres.
DIMALTU	2	Alternate units
DIMALTZ	0	Alternate unit zero suppres.
DIMAPOST	2	Prefix and suffix for alt. text
DIMASO	ON	Create associative dimens.
DIMASZ	.125	Arrow size
DIMAUNIT	0	Angular unit format
DIMBLK	NONE	Arrow block
DIMBLK1	NONE	Separate arrow block 1
DIMBLK2	NONE	Separate arrow block 2
DIMCEN	–.05	Center mark size
DIMCLRD	BYBLOCK	Dimension line color
DIMCLRE	BYBLOCK	Extension line color
DIMCLRT	BYBLOCK	Dimension text color
DIMDEC	BYBLOCK	Decimal places
DIMDLE	0	Dimension line extension
DIMDLI	.38	Dimension line increment
DIMEXE	.12	Extension line extension
DIMEXO	.06	Extension line offset
DIMFIT	3	Fit text
DIMGAP	.09	Dimension line gap
DIMJUST	0	Justif. of txt on dim. line
DIMLFAC	1.00	Linear unit scale factor
DIMLIM	OFF	Generate dimension limits
DIMPOST		Prefix & suffix for dim. text

Figure 1.157 Type DIMSTYLE at the Command line to obtain a listing of the dimension variables, their settings, and their definitions.

DIMENSIONING SYSTEM VARIABLES

DIM VARS	DEFAULT	DEFINITION
DIMRND	0	Rounding value
DIMSAH	OFF	Separate arrow blocks
DIMSCALE	1	Dimension scale factor
DIMSD1	OFF	Suppress 1st dimension line
DIMSD2	OFF	Suppress 2nd dimension line
DIMSE1	OFF	Suppress 1st extension line
DIMSE2	OFF	Suppress 2nd extension line
DIMSHO	OFF	Show dragged dimensions
DIMSOXD	OFF	Suppress outside dim. lines
DIMSTYLE	UNNAMED	Dimension style
DIMTAD	OFF	Place text above dim. line
DIMTDEC	4	Tolerance decimal places
DIMTFAC	1	Tolerance text scale factor
DIMTIH	ON	Text inside horizontal
DIMTIX	OFF	Text inside extension lines
DIMTM	0	Minus tolerance value
DIMTOFL	OFF	Tolerance dimensioning
DIMTOH	ON	Text outside horizontal
DIMTOL	OFF	Tolerance dimensioning
DIMTOLJ	0	Tolerance vert. justification
DIMTP	1	Plus tolerance value
DIMTSZ	0	Tick size
DIMTVP	0	Text vertical position
DIMTXSTY	STAND.	Text style
DIMTXT	.125	Text size
DIMTZIN	0	Tolerance zero suppression
DIMUNIT	2	Unit format
DIMUPT	OFF	User positioned text
DIMZIN	0	Zero suppression

Figure 1.158 Additional dimension variables are shown here as a continuation of Fig. 1.157.

1.55 DIMSTYLE Variables

Type DIMSTYLE at the Command line, STATUS, and press (ENTER) to obtain a screen of dimensional system variables, their assigned values, and their definitions **(Fig. 1.155)**. Sizes of dimensioning variables are based on the letter height, which is most often 0.125" **(Fig. 1.156)**.

To set and save a few variables needed for basic applications, OPEN the file PROTO1. Each variable is set by typing SETVAR and the name of the **dimensioning variable** (DIMTXT, text height, for example) and assigning a numerical value. A list of the 58 dimensioning variables is given in **Figs. 1.157** and **1.158**. Assign the basic variable values of DIMTXT, DIMASZ, DIMEXE, DIMEXO, DIMTAD, DIMDLI, DIMASO, and DIMSCALE (Fig. 1.156) to PROTO1, since they apply to most applications. Type UNITS and set decimal fractions to two decimal places for inches.

SAVE these settings to PROTO1 with no objects on it and use it as the prototype when creating a new file (DWG-3, for example), which becomes the current file with the same settings as PROTO1.

Dimensioning variables can be set from dialog boxes also, instead of being typed, and these techniques are covered later. You will be more proficient by becoming familiar with both methods of

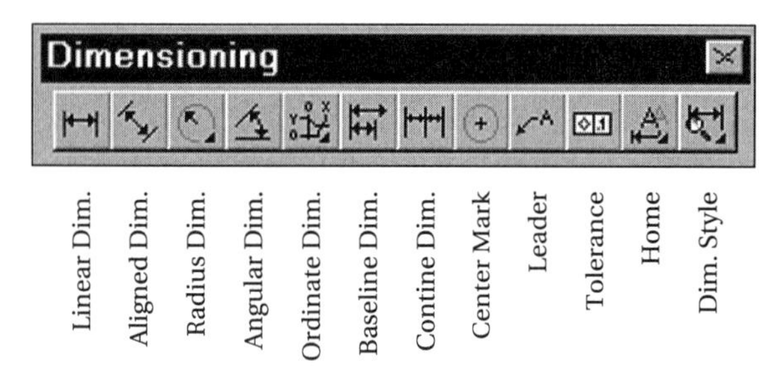

Figure 1.159 Dimensioning options from the `Dimensioning` toolbar are shown here.

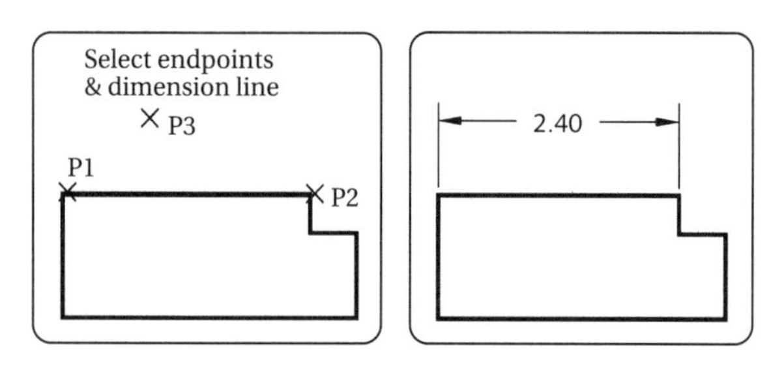

Figure 1.160 Dimensioning a line.

Step 1 `Command:` <u>`DIMLINEAR`</u>

`First extension line origin or` (ENTER) `to select:` <u>`P1`</u>

`Second extension line origin:` <u>`P2`</u>

`Dimension line location (Text/Angle/Horizontal/Vertical/Rotated) :` <u>`P3`</u>

`Dimension text <2.40>:` (ENTER)

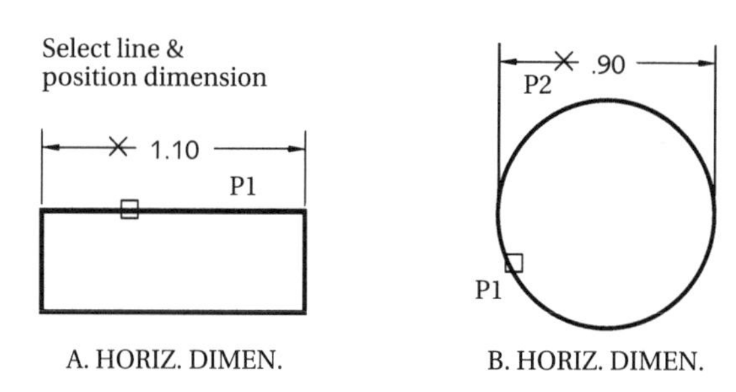

Figure 1.161 Semiautomatic dimensioning of a line. Specify a horizontal dimension and when prompted for the `First extension origin`, press (ENTER) to get the prompt, `Select entity`. Select the object to be measured and dimensioned and locate the position of the dimension line. The results of dimensioning a line and a circle are shown at **A**

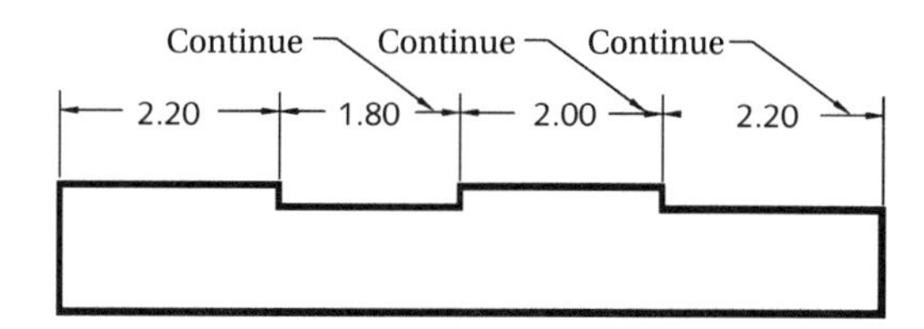

Figure 1.162 When dimensions are placed end to end, the `Continue` option is used to specify the second extension line origin after the first dimension line has been drawn.

assigning variables. For now, use the `DWG-3` file to explore the fundamentals of dimensioning.

1.56 DIMLINEAR (Dimensioning Toolbar)

The `Dimensioning` toolbar **(Fig. 1.159)** is a convenient means of selecting dimensioning commands. Select the `Linear Dimension` icon, the points as prompted, and the horizontal option as shown in **Fig. 1.160**.

A dimension is applied semiautomatically in **Fig. 1.161** by pressing (ENTER) when prompted for `Endpoints`, selecting the line or circle to be dimensioned, and locating its dimension line.

The `DIMCONTINUE` command is selected from `Dimensioning` toolbar (or typed at the `Command` line) to continue a chain of linear, angular, or ordinate dimensions from the last extension line **(Fig. 1.162)**. `Baseline` applies dimensions from a single endpoint and each dimension incrementally offset by the dimension line increment variable, `DIMDLI` **(Fig. 1.163)**.

BASE-LINE DIMENSIONS

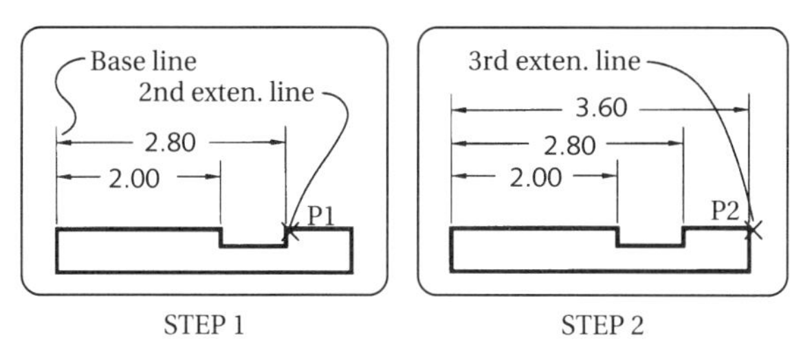

Figure 1.163 `Baseline` option.
Step 1 `Command:` <u>`DIM`</u> (ENTER)
Dimension the first line as shown in Fig. 1.160. When prompted for the next dimension, type `BASELINE` (ENTER) and you will be prompted for the `Second extension line origin:` <u>`P1`</u>.
Step 2 The second dimension line will be drawn. Respond to `Dim:` with `BASELINE` (ENTER) again, and select <u>`P2`</u> when asked for `Second extension line origin.` Continue in this manner for other dimensions.

LINEAR DIMENSIONS

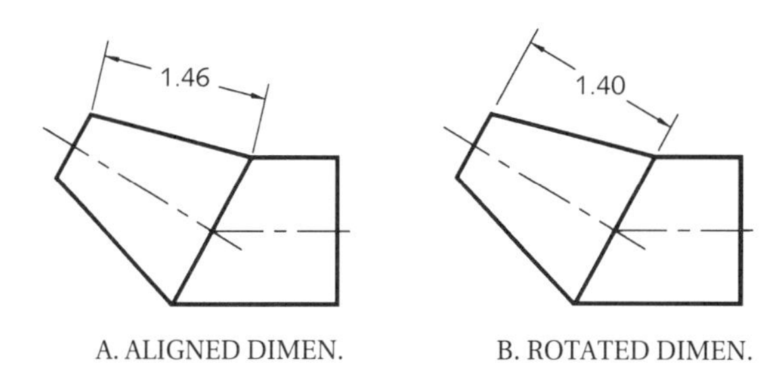

Figure 1.164 Linear dimensions can be `Aligned` (**A**) by selecting any two points, or `Rotated` (**B**) by specifying an angle for the dimension line.

Select the `DIMALIGNED` command from the `Dimensioning` toolbar and you will be prompted to select the `1st` and `2nd` extensions lines and the position of the dimension line **(Fig. 1.164A)**. The dimension line will be inserted aligned with line 1-2. Extension lines can be automatically drawn by pressing (ENTER) at the first prompt and selecting the line to be dimensioned or rotated to a specified angle using the `Rotated` dimension line location option **(Fig. 1.164B)**.

DIMENSIONING ANGLES

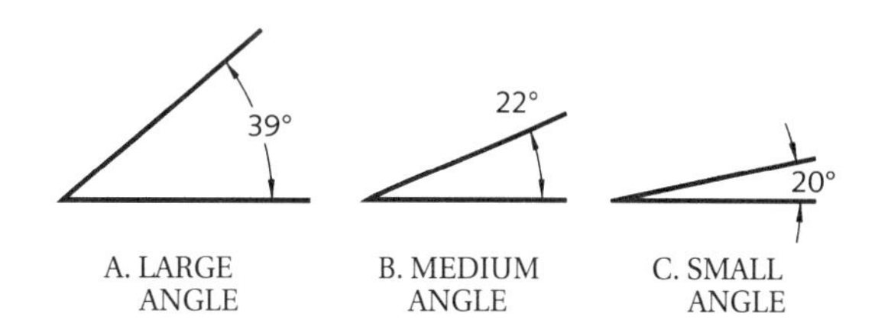

Figure 1.165 Angles will be dimensioned in any of these three formats using AutoCAD.

DIMENSIONING AN ANGLE

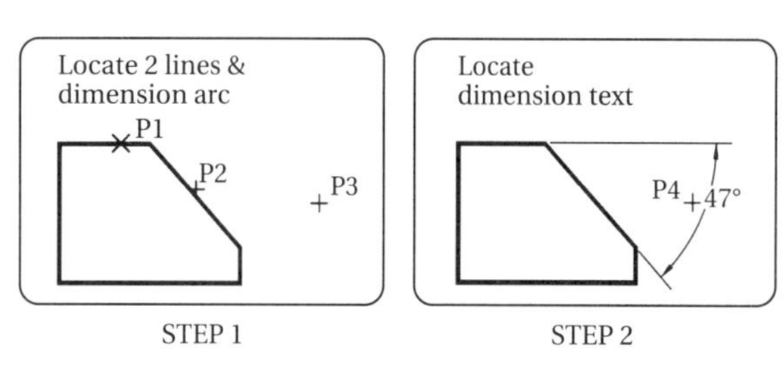

Figure 1.166 `DIMANGULAR` command.
Step 1 `Command:` <u>`DIMANGULAR`</u>
`Select arc, circle, line, or RETURN:` <u>`P1`</u>
`Second line:` <u>`P2`</u>
`Dimension arc line location (Text/Angle):` <u>`P3`</u>
`Dimension text <47°>:`
Step 2 Enter text location: <u>P4</u> or (ENTER)

1.57 DIMANGULAR (Dimensioning Toolbar)

Figure 1.165 shows variations for dimensioning angles depending on the space available. Select the `DIMANGULAR` icon, select lines of the angle, and locate the dimension line arc as shown in step 1 of **Fig. 1.166**.

An angle dimension can be applied by selecting its vertex and the endpoints of each line as shown in step 1 of **Fig. 1.167**. For angles over 180°, select a point on the sides of the angle and move counterclockwise to the second point (step 2 of Fig. 1.167) and locate the dimension arc.

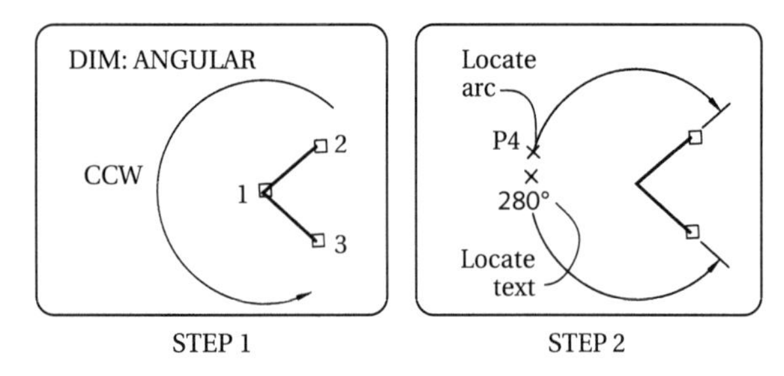

Figure 1.167 `DIMANGUALR` command using endpoints for angles over 180°.
Step 1 `Command:` <u>`DIMANGULAR`</u> (ENTER)
`Select arc, circle, line, or RETURN:`
(ENTER)
`Angle vertex:` <u>`P1`</u>
`First angle endpoint:` <u>`P2`</u>
`Second line:` <u>`P3`</u>
Step 2 `Dimension arc line location (Text/Angle):` <u>`P4`</u>
`Dimension text <280°>:`

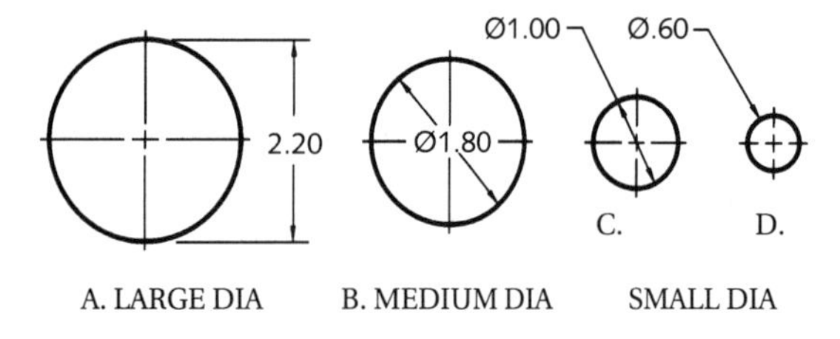

Figure 1.168 Types of dimensions available for dimensioning circles.

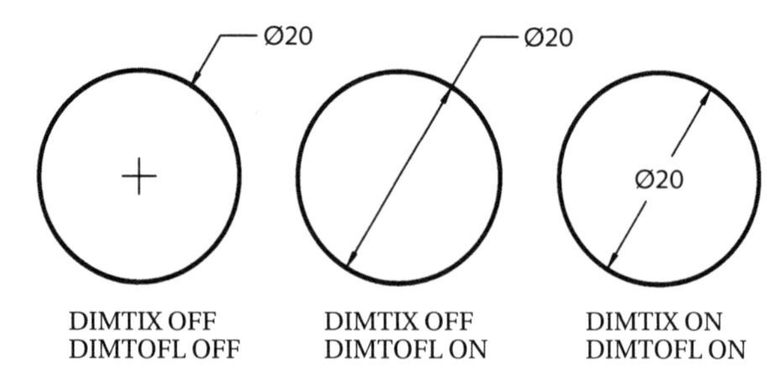

Figure 1.169 Examples of dimensions with various `DIM VARS` settings.

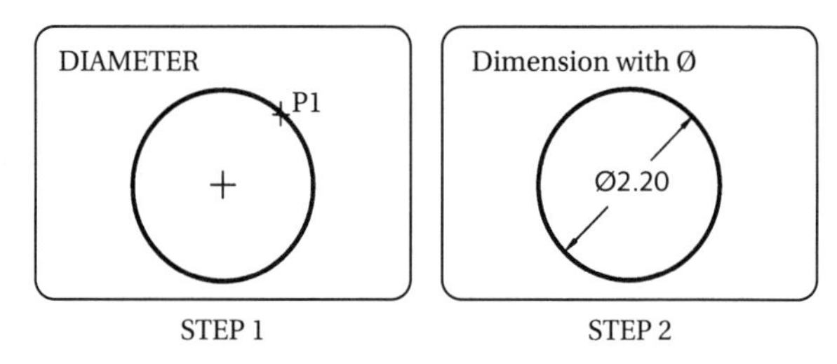

Figure 1.170 Dimensioning a circle.
Step 1 `Command:` <u>`DIMDIAMETER`</u> (ENTER)
`Select arc or circle:` <u>`P1`</u>
`Dimension text <2.20>:`
Step 2 `Dimension line location (Text/Angle):` (ENTER) (Accepts 2.20 and the dimension is drawn from `P1` through the center.)

1.58 DIMDIAMETER (Dimensioning Toolbar)

Diameters of circles can be placed as shown in **Fig. 1.168** depending on the available space. By changing system variables `DIMTIX` and `DIMTOFL`, circles can be dimensioned as shown in **Figs. 1.169** and **1.170**.

By setting the `DIMTIX` system variable `On`, the text is forced inside the extension lines regardless of the available space. The `DIMTOFL` (`On`) system variable forces a dimension line to be drawn between extension lines when the text is located outside.

1.59 DIMRADIUS (Dimensioning Toolbar)

Select `DIMRADIUS` from the `Dimensioning` toolbar to dimension areas with an `R` placed in front of the text (`R1.00`, for example) as shown in **Fig. 1.171**. For horizontal dimension text, if the angle of the radius dimension line is greater than 15 degrees from horizontal and the dimension won't fit with-

DIMENSIONING ARCS

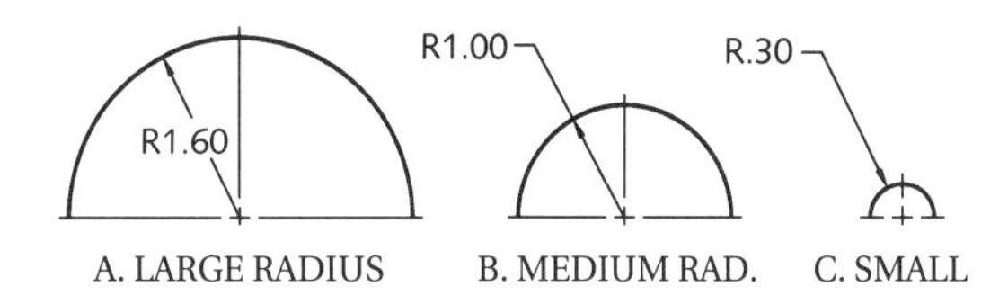

Figure 1.171 Arcs will be dimensioned by one of the formats given here depending on the size of the radius.

DIM: RADII

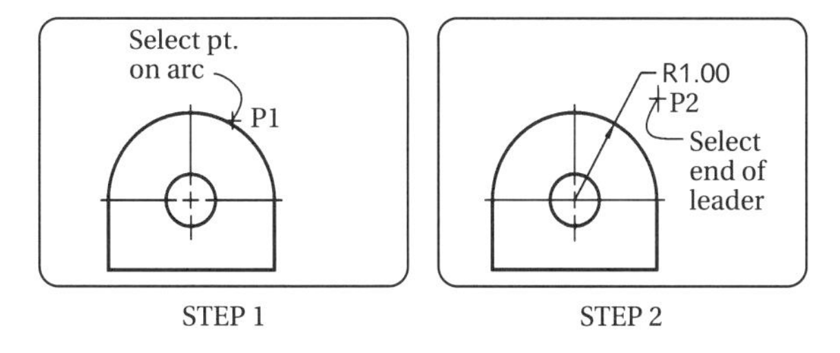

Figure 1.172 Radius drawn with leader.
Step 1 `Command:` <u>`DIMRADIUS`</u>
`Select arc or circle:` <u>`P1`</u>
`Dimension text <1.00>:` (ENTER)
`Dimension line location (Text/Angle):` <u>`P2`</u>
(The leader and `R1.00` are drawn.)

in the radius, a leader line is drawn along with the dimension **(Fig. 1.172).**

The `LEADER` command is used to add a line that connects an annotation to a feature. It can be composed of an arrowhead attached to curved or straight line segments. The annotation must be known and typed in **(Fig. 1.173)**.

1.60 DIMSTYLE Variables (Dimensioning Toolbar)

Select `Dimension Style...` (`DDIM`) from the `Data` menu to display the `Dimension Styles` dialog box shown in **Fig. 1.174**, from which a number of variables can be assigned for three major categories: `Geometry`, `Format`, and `Annotation`. Each group of settings can be made and saved by name (`NEW-4`, for example) for future use.

LEADER OPTION

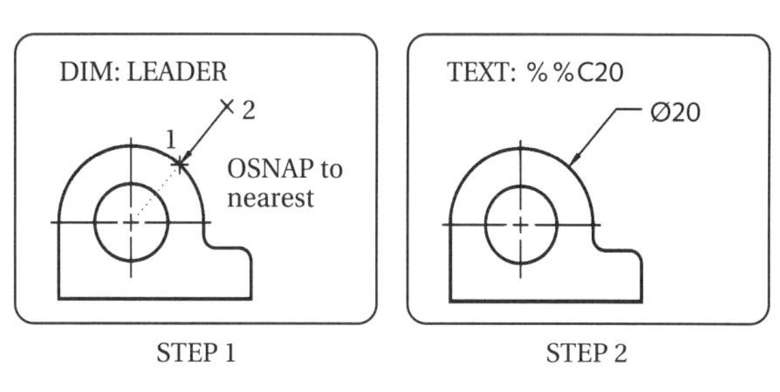

Figure 1.173 Using the `LEADER` command.
Step 1 `Command:` <u>`LEADER`</u>
`From point: (OSNAP to NEAREST)`
`To point:` <u>`P2`</u>
`To point (Format/Annotation/Undo)`
`<Annotation>:` (ENTER)
Step 2 `Annotation (or RETURN for options):` <u>`%%C20`</u>
`Mtext:` (ENTER)

DIMENSION STYLES (DDIM)

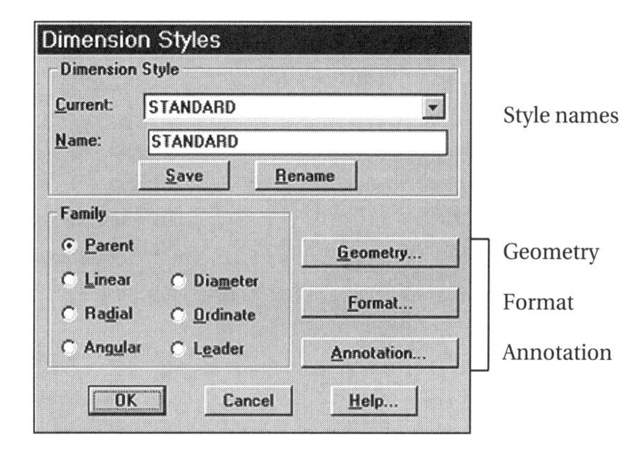

Figure 1.174 At the `Command` line, type <u>`DDIM`</u> to obtain this menu box with three major submenus: `Geometry`, `Format`, and `Annotation`.

Geometry

Click on the `Geometry...` button to display the `Geometry` dialog box shown in **Fig. 1.175**. The options under the `Dimension-Line` group controls the following variables: `DIMSD1`, `DIMSD2`, `DIMDLE`, `DIMDLI`, and `DIMCLRD`. The `Suppress 1st` and `2nd` boxes turn `On` the `DIMSD1` and

GEOMETRY (DDIM/DIM STYLE)

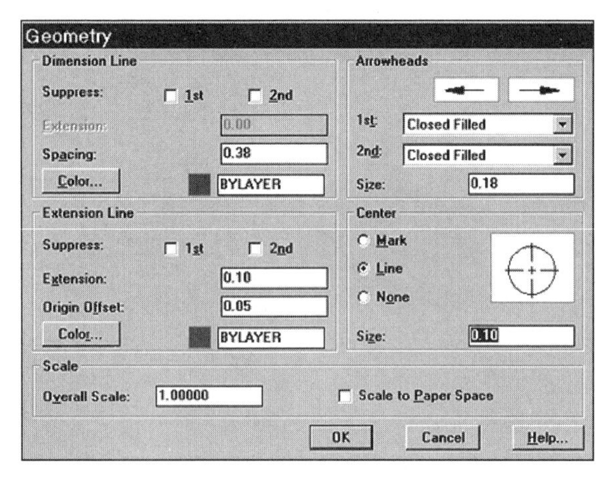

Figure 1.175 The Geometry box (opened from the Dimension Styles menu) offers a number of dimensioning settings.

DIMSD2 variables to suppress the first and second dimension lines. When Oblique-Stroke arrows are used, the value placed in the Extension box specifies the distance the dimension line extends beyond the extension line. The Spacing box is used to set DIMDLI which controls the spacing between baseline dimensions. The Color button displays the color menu from which to select a color for the dimension line (DIMCLRD).

The options of the Extension-Line area (Fig. 1.175) control the following variables: DIMSE1, DIMSE2, DIMEXE, DIMEXO, and DIMCLRE. The Suppress 1st and 2nd boxes turn On the DIMSE1 and DIMSE2 variables to suppress the first and second extension lines. The value typed in the Extension box specifies the distance the extension line extends beyond the dimensioning arrow (DIMEXE). The Origin Offset option is used to specify the size of the gap between the object and the extension line (DIMEXO). The Color button lets you select the color of the extension lines (DIMCLRE).

The options of the Arrowheads area **(Fig. 1.176)** control the following variables: DIMASZ, DIMTSZ, DIMBLK1, and DIMBLK2. The value typed

ARROWHEAD AREA (GEOMETRY BOX)

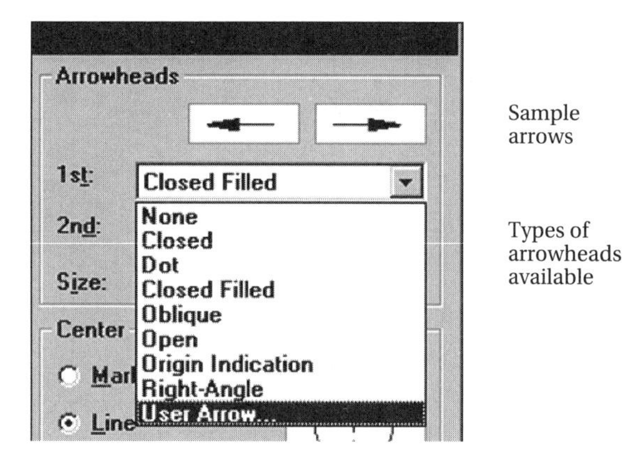

Figure 1.176 From the Geometry menu, Arrowheads can be selected from the list.

in the Size box gives the size of the arrowhead, the DIMASZ variable. By selecting arrows next to the 1st or 2nd boxes, the types of arrowheads for each end of the dimension are displayed in a drop-down list (Fig. 1.176). If only the 1st arrow type is selected, it is automatically applied to the second end unless a 2nd arrow type is specified. Tick marks are used when Oblique arrows are selected, in which case the Extension edit box becomes active so you can specify the DIMTIX variable value. When User Arrow is selected, custom-made arrows can be inserted (DIMBLK1 and DIMBLK2).

Select Mark, Line, or None (equivalent to the DIMCENTER command) to draw center marks, center lines, or nothing on arcs and circles when diameter or radius dimensions are placed outside by the DIMDIAMETER and DIMRADIUS commands. The value placed in the Size box gives the size of the center mark (a plus mark). A minus value gives center lines and a zero gives none.

The Overall Scale box specifies DIMSCALE, which controls the geometry of dimensioning variables such as arrow size, text height, extension-line offsets, center size, and others. When set to 1, these variables are full size, and when set to 2, they are twice as big. The Scale to Paper Space box (Fig. 1.175) is picked to adjust

FORMAT BOX (DDIM)

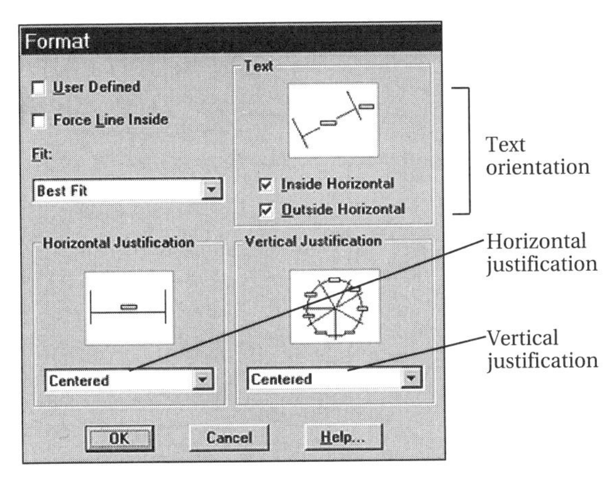

Figure 1.177 The `Format` box (opened from the `Dimension Styles` box) offers a number of settings for the placement of text.

BEST FIT EXAMPLES

70 — All inside
40 — Arrows inside
50 — Text inside
20 — All outside

Figure 1.178 Examples of dimensions applied using the `Best Fit` option.

the dimensions in model space to the scale of paper space. A default of `1` is used if you are not working in paper space.

Format

Select the `Format` button (Fig. 1.174) to display the `Format` dialog box in **Fig. 1.177** to control the placement of text, arrows, leaders, and dimension lines. Select `User Defined` (`DIMUPT`) to place the text of a dimension with the pointer, and disable this option for the text to be placed automatically. Select the arrow next to the `Fit` edit box (`DIMFIT`) to display a drop-down list from which to pick the options of `Text and Arrows`, `Text Only`, `Arrows Only`, `Best Fit`, `Leader`, and `No Leader` to determine what is placed inside extension lines if space is limited **(Fig. 1.178)**.

When `Inside Horizontal` (`DIMTIH`) and `Outside Horizontal` (`DIMTOH`) are picked, text inside extension lines is placed horizontally and text outside extension lines is placed horizontally **(Fig. 1.179)**.

Select the arrow under `Horizontal Justification` to display the `Centered`, `1st Extension Line`, `2nd Extension Line`, `Over 1st Extension`, and `Over 2nd Extension` options for placing text along the dimension line as shown in **Fig. 1.180**. `Vertical Justification` has options of `Centered`, `Above`, `Outside`, and `JIS` (Japanese Industrial Standard) for positioning text relative to the dimension line **(Fig. 1.181)**.

DIMTOH AND DIMTIH VARIABLES

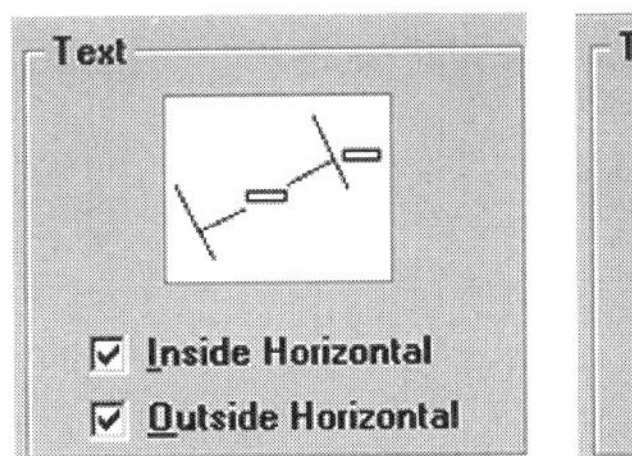

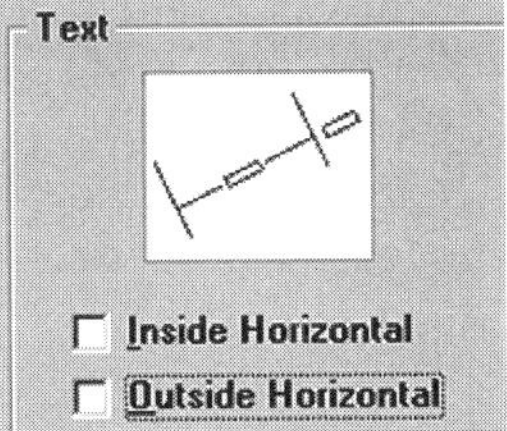

Figure 1.179 The examples show the results of having the `DIMTOH` and `DIMTIH` variables set to `On` and to `Off`.

HORIZONTAL JUSTIFICATION

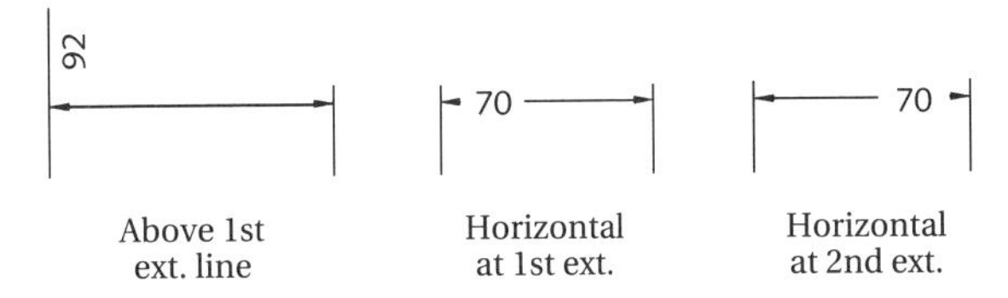

Figure 1.180 Examples of different settings of `Horizontal Justification`.

Annotation

Click on the `Annotation` button (Fig. 1.174) to display the dialog box shown in **Fig. 1.182,** which

VERTICAL JUSTIFICATION

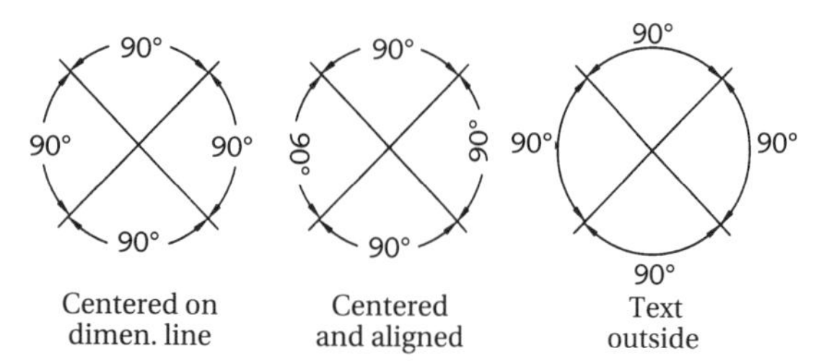

Figure 1.181 Examples of different settings of `Vertical Justification`.

ANNOTATION (DDIM\DIM STYLE)

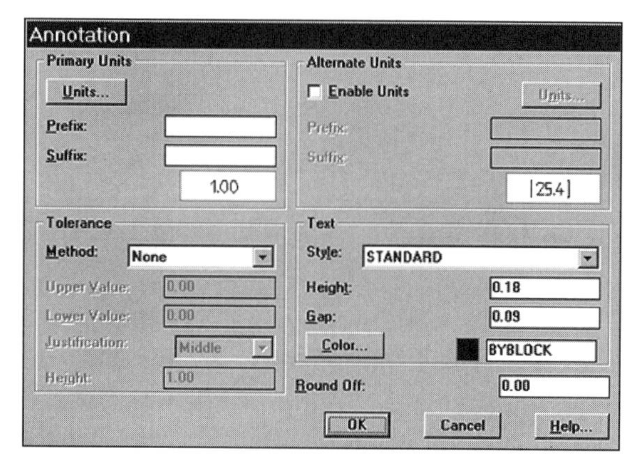

Figure 1.182 The `Annotation` box (opened from the `Dimension Styles` box) offers a number of settings for the type of text used in dimensioning.

PRIMARY UNITS

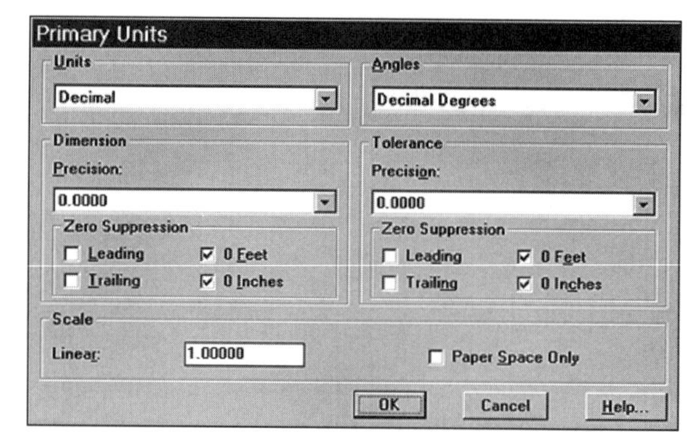

Figure 1.183 From `Units` of the `Annotation` box, this `Primary Units` box is displayed for making settings for dimensioning text.

SUPPRESS ZEROS-ARCH. UNITS (DIMZIN)

0 Ft 0 In	1/4"	4"	1'	1'–01/4"
No options	0'–01/4"	0'–6"	2'–0"	1'–01/4"
0 In	0'–01/4"	0'–4"	1'	1'–01/4"
0 Ft	1/2"	4"	1'–0"	1'–01/4"

Figure 1.184 The `Zero Suppression (DIMZIN)` options control the leading and trailing zeros in dimensioning, especially when applied to architectural dimensions.

controls the actual dimension text. Selecting the `Units...` button in the `Primary Units` area displays the `Primary Units` dialog box **(Fig. 1.183)**. Select the type of units (decimal, engineering, and so on) from the `Units` drop-down list. Use the `Dimension Precision` drop-down list to select the number of decimal places (or fractional accuracy) desired. The type of angle measurement is selected from the `Angles` drop-down list and their precision can similarly be specified from the `Tolerance Precision` drop-down list.

Check boxes control the `DIMZIN` value, suppressing zero values before (leading) or after (trailing) the decimal point. For example, select both `Leading` and `0 Inches` to make 0.20 become .20. **Figure 1.184** shows the results of applying the four options to architectural units. The `Linear` value in the `Scale` area controls the `DIMLFAC` value. Select `OK` to return to the `Annotation` dialog box.

Under `Alternate Units`, select the `Enable Units` check box for dual dimensions (`DIMALT`) to be used in dimensions; the scale of 25.4 gives millimeter equivalents for inches as an alternate dimension. When the `Units...` button is selected, the `Alternate Units` dialog box is displayed **(Fig. 1.185)** and settings are made in the same way

ALTERNATE UNITS (DUAL DIMENSIONING)

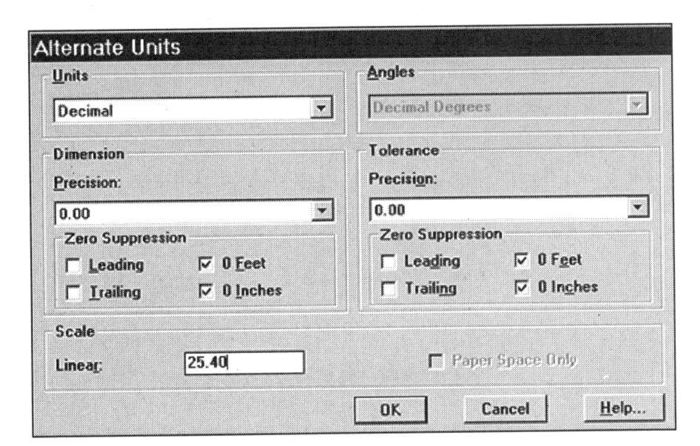

Figure 1.185 Select Alternate Units from the Annotation box (Fig. 1.182) to obtain this box for making specifications for alternate units used in dual dimensioning.

ALTERNATE UNITS (DIMALT)

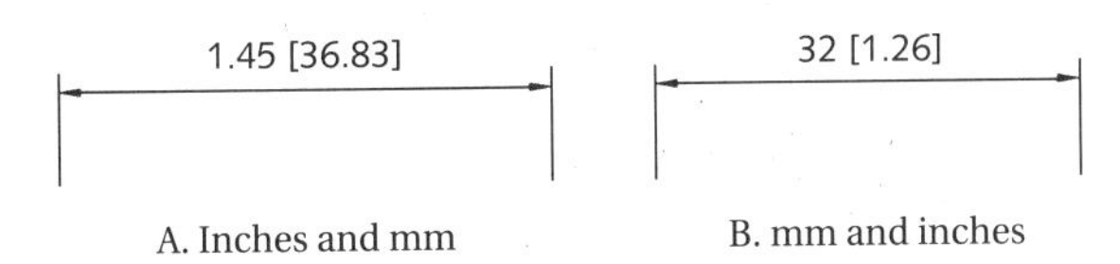

A. Inches and mm

B. mm and inches

Figure 1.186 Examples of alternate units (dual dimensions) made in inches and millimeters.

as Primary Units. Examples of dimensions with alternate units are shown in **Fig. 1.186**.

From the Annotation dialog box a Prefix and Suffix can be specified for both the Primary (DIMPOST) and Alternate Units (DIMAPOST) **(Fig. 1.187)**. A Tolerance Method of None, Symmetrical, Deviation, Limits, and Basic can be selected from the drop-down list and an Upper (DIMTP) and Lower Value (DIMTM), Justification (DIMTOLJ) and Height assigned, examples of which are shown in **Fig. 1.188**. Height is stored to the DIMTFAC variable as a ratio of the tolerance height to the main dimension text height.

The Text area in Fig. 1.182 gives options for selecting a text Style (DIMTXSTY) created with the STYLE command, assigning a text Height (DIMTXT), specifying the Gap (DIMGAP) around the

TOLERANCE AREA (ANNOTATION BOX)

Tolerance

Method:	Deviation	Type of tolerance
Upper Value:	0.0020	Limits of tolerance
Lower Value:	0.0030	
Justification:	Middle	Justification
Height:	1.0000	Tolerance text height

Figure 1.187 From the Tolerance area of the Annotation box (Fig. 1.182), the format for tolerances can be specified.

DIMENSION TOLERANCE FORMATS

2.0000±.0020	2.0000 +.0030 −.0020	2.0030 1.9980
DIMTP & DIMTM SAME	DIM & DIMTM DIFFERENT	DIMLIM

Figure 1.188 Examples of various formats for tolerancing are shown here.

SAVING DIMENSIONING STYLES

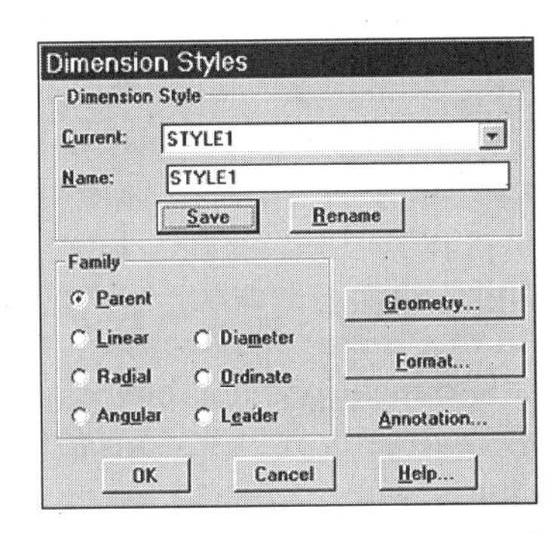

Figure 1.189 From the Dimensioning toolbar, select Dimension Style, Parent, and Save to keep this setting.

text, and picking a Color for the text. Set the Round Off value (DIMRND) for all dimensions except angular measurements. Press the OK button to return to the Dimension Style dialog box **(Fig. 1.189)**.

1.61 Saving Dimension Styles

After setting the previous dimensioning variables, they can be saved as a parent dimension style (DIMSTYLE) by picking the Parent radio button in the Dimension Styles dialog box (Fig. 1.189), typing the name (STYLE-1 for example) in the Name: edit box, and picking the Save button.

Style families can be created to include special applications of variables. If you want to Force Line Inside (DIMTOFL) for diameter dimensions only, select the family name button Diameter, select the Parent Style from the Current Style list, make modifications to the Parent, and Save. When this Parent Style is opened, AutoCAD applies the family member style automatically.

1.62 Dimension Style Override (Dimensioning Toolbar)

If you want to Override (DIMOVERRIDE) one or more settings previously saved to a Dimension Style, from the Dimensioning toolbar, select Dimension Style, and pick the Style from the Current Style list. If you want to change DIMSCALE, for example, make this change to the style and choose the OK box. The style name is listed with a plus mark, +STYLE-1, for example.

To apply the Override settings, type DIMSTYLE at the Command line, enter A for APPLY, and select the dimensions that are to be updated **(Fig. 1.190)**. To RESTORE dimensions to a different set of variables, type DIMSTYLE at the Command line, enter R for RESTORE, and select the dimensions whose style is to be restored.

1.63 DIMEDIT (Command Line)

Type DIMEDIT at the Command line to obtain the options of Home, New, Rotate, and Oblique for

DIMSCALE OVERRIDE

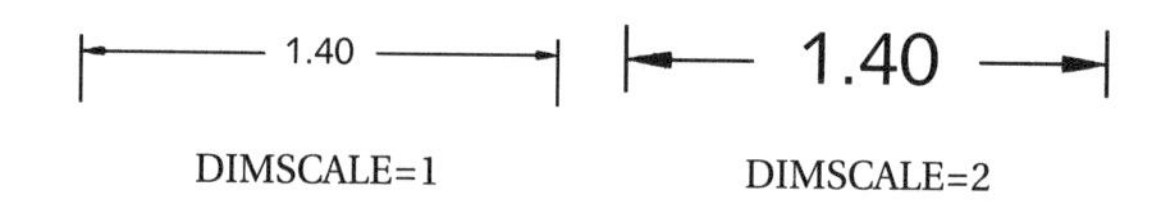

Figure 1.190 When DIMSCALE is changed from 1 to 2 and OVERRIDE used, the selected dimension is updated.

DIMASO: HOMETEXT

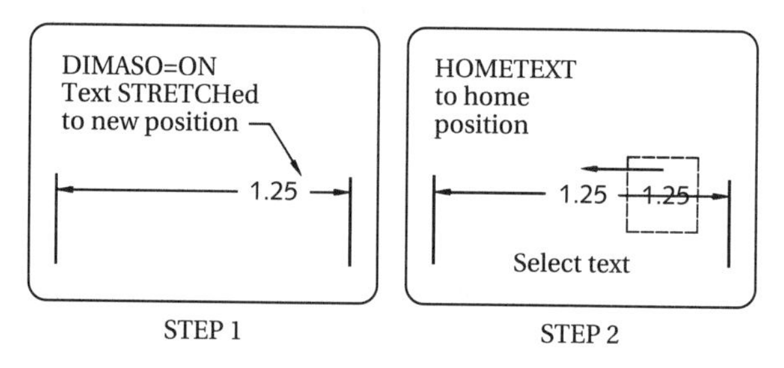

Figure 1.191 HOMETEXT option.
Step 1 Turn DIMASO on and dimension the object. If the object and dimensions are STRETCHed, the dimension numerals will not be centered.
Step 2 Dim: HOMETEXT (ENTER)
Select objects: (Select text.) (The text will automatically center itself in the dimension line.)

DIMASO: NEWTEXT

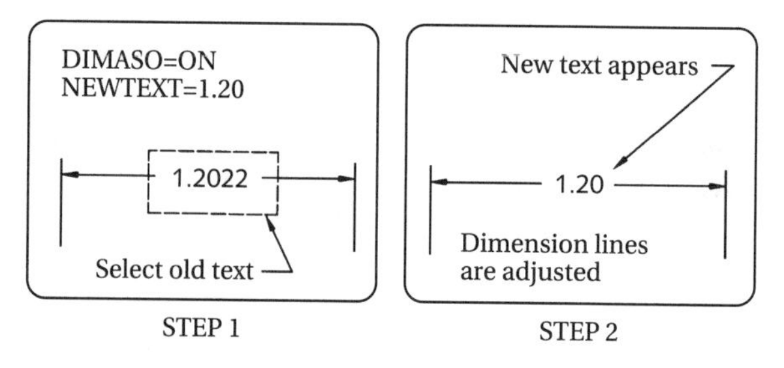

Figure 1.192 NEWTEXT option.
Step 1 Turn DIMASO on and dimension the object.
Dim: NEWTEXT (ENTER)
Enter new dimension text: 1.20 (ENTER)
Select objects: (Select text to be changed.)
Step 2 The old text will be replaced with new text.

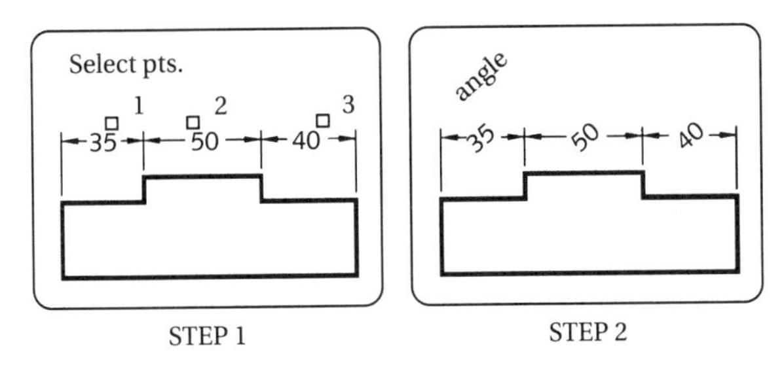

Figure 1.193 `TROTATE` option.
Step 1 `Command:` `DIM` (ENTER)
`Dim:` `TROTATE`
`Enter new text angle:` `45`
`Select objects:` (Select each associative dimension.)
Step 2 Press (ENTER) and the text is redrawn at an angle within the dimension lines.

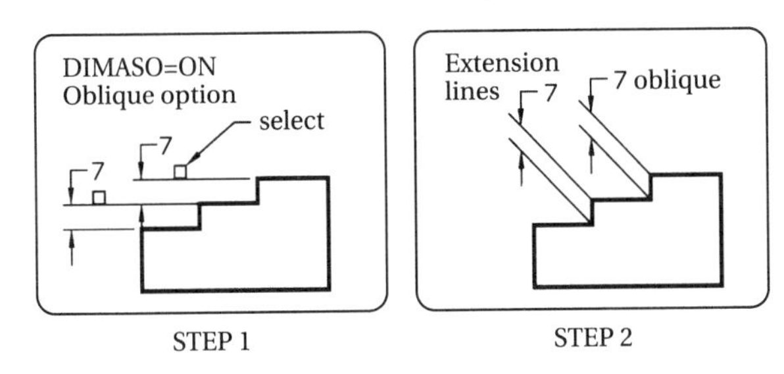

Figure 1.194 `OBLIQUE` option.
Step 1 `Command:` `DIM` (ENTER)
`Dim:` `OBLIQUE`
`Select objects:` (Select an associative dimension.)
`2 selected, 2 found`
`Select objects:` (ENTER)
`Enter obliquing angle (RETURN for none):` `45`
Step 2 Press (ENTER) and the extension lines are drawn at an angle and the dimension numerals are repositioned.

changing existing dimensions. `Home` repositions text to its standard position at the center of the dimension line after being changed by the `STRETCH` commands **(Fig. 1.191)**. `New` changes text within a dimension line **(Fig. 1.192)** by press-

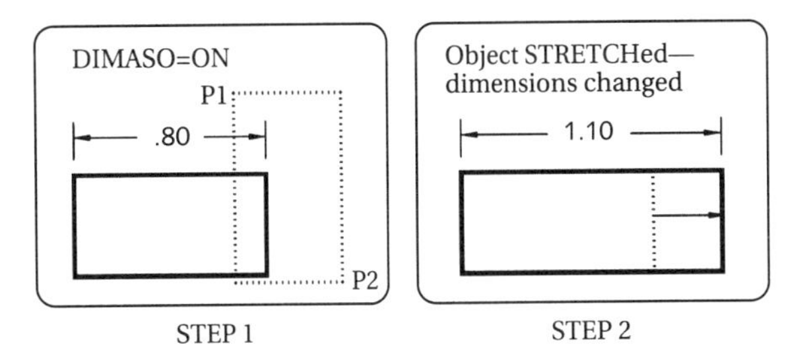

Figure 1.195 Associative dimensions.
Step 1 Set `DIMASO` on and apply dimensions to the part. Use the `STRETCH` command and `CROSSING WINDOW` at the end of the part.
Step 2 Select a new endpoint for the part and it will be lengthened and new dimensions will be calculated.

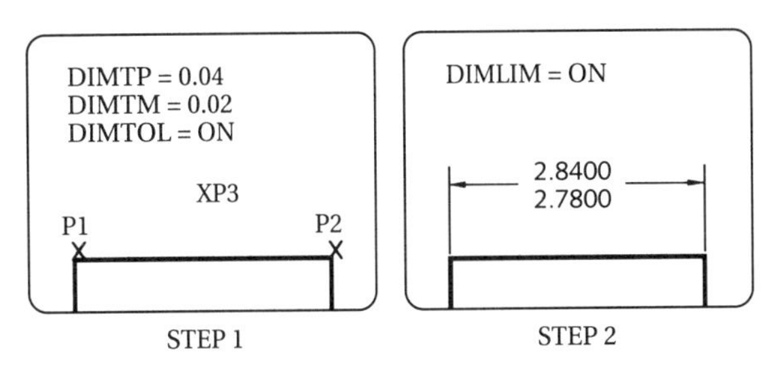

Figure 1.196 Dimensions can be toleranced in any of these three formats.

ing (ENTER) when prompted and inserting the new text. `Rotate` positions dimension text at any specified angle **(Fig. 1.193)**. `Oblique` creates oblique extension lines **(Fig. 1.194)**.

1.64 Stretching Dimensions

Dimensioned parts can be changed in size with the `STRETCH` command while associated dimensions are automatically changed as they are `STRETCH`ed **(Fig. 1.195)**. If the variable `DIMASHO` is set to `1` (`On`), new dimension text is seen being dynamically changed as the dimensions are stretched.

GEOMETRIC TOLERANCE FEATURE CONTROL FRAME

⌖	Ø0.025 Ⓜ	A Ⓜ	B Ⓢ	C Ⓛ

True Pos. / DIA symbol / Tolerance / Max Matl. / Primary / Max Matl. / Secondary / Regardless of size / Tertiary / Lease Matl.

Figure 1.197 The parts of a feature control frame that give geometric tolerance specifications are defined here.

GEOMETRIC TOLERANCE SYMBOLS

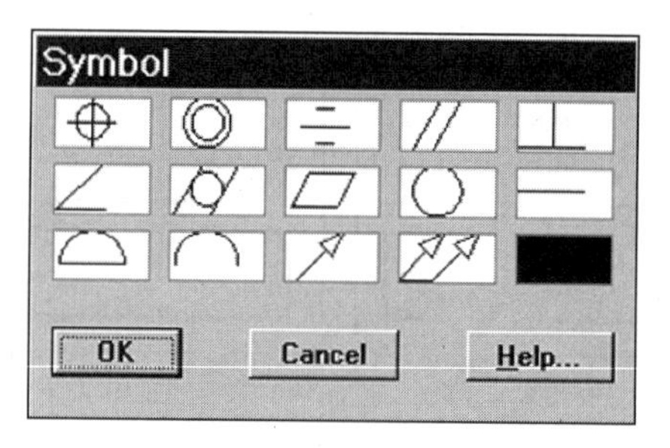

Figure 1.198 The symbols of geometric tolerance are found in the `Symbol` box.

1.65 Toleranced Dimensions

Dimensions can be toleranced automatically using the `Dimension Styles` defined previously in which the settings shown in **Fig. 1.196** were made: `DIMTOL` (tolerance on), `DIMTP` (plus tolerance), and `DIMTM` (minus tolerance). When `DIMLIM` is `On`, the upper and lower limits of the size are given.

`DIMTFAC` is a scale factor that controls the text height of the tolerance values, which is about 80% of the basic dimension.

1.66 Geometric Tolerances (Dimensioning Toolbar)

Geometric tolerances specify the permissible variations in form, profile, orientation, location, and runout. A typical geometric tolerance feature control frame is given in **Fig. 1.197**.

From the `Dimensioning` toolbar, select `Tol-erance` to display the `Symbol` dialog box **(Fig. 1.198)**, select the desired symbol, and pick `OK` to display the `Geometric Tolerance` box **(Fig. 1.199)**. In `Tolerance 1`, select `Dia` to insert the symbol; under `Value`, type the first tolerance `Value`, pick `MC` to give the material condition; and pick the material condition from the `Material Condition` dialog box. Give the `Tolerance 2` value (optional) in the same way as `Tolerance 1`.

Under `Datum 1`, enter the letter for the primary reference datum, choose `MC`, and select your preference from the `Material Condition` box. Add `Datum 2` and `Datum 3` values in the same manner, select `OK`, and pick the location of the feature control frame **(Fig. 1.200)**.

1.67 Digitizing with the Tablet

Drawings on paper can be taped to a digitizing tablet and digitized point-by-point. A drawing is calibrated in the following manner:

`Command:` <u>`TABLET`</u> (ENTER)

`Option (ON/OFF/CAL/CFG):` <u>`CAL`</u> (ENTER)

(Calibrate tablet for use.)

`Digitize first known point:` (Digitize point.)

`Enter coordinates for first point:` <u>`1,1`</u> (ENTER)

`Digitize second known point:` (Digitize point.)

`Enter coordinates for second point:` <u>`10,1`</u>

Digitize points from left to right, or from bottom to top of the drawing. Use `On` or `Off` to turn the

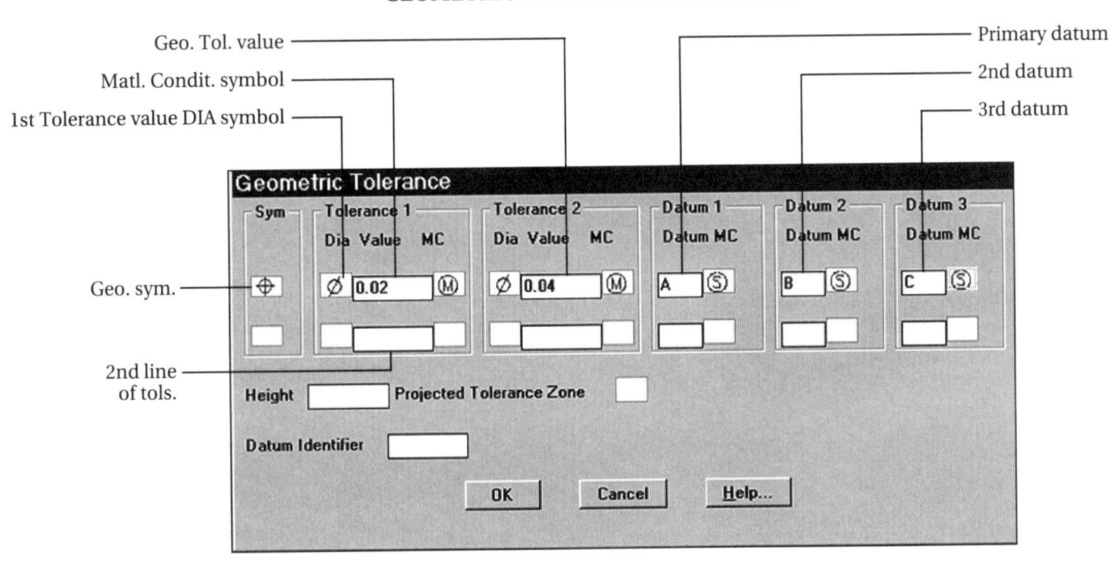

Figure 1.199 The `Geometric Tolerance` box, found under `Tolerance` of the `Dimensioning` toolbar, is used to give specifications for geometric tolerances.

GEOMETRIC TOLERANCE FEATURE CONTROL FRAMES

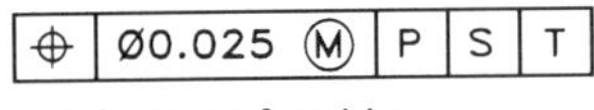

A. Tolerance of position

⊥	0.004	A

B. Tolerance of perpendicularity

Figure 1.200 Applications of `Geometric Tolerances` in feature control frames are shown here.

tablet mode on or off. Function key (F4) also turns the tablet off so the cursor can select from the screen menus. To draw lines, select the `LINE` command from the `Draw` toolbar (or type `L`), and pick points on the tablet with the stylus.

1.68 SKETCH (Miscellaneous Toolbar)

The `SKETCH` command can be used with the tablet for tracing drawings composed of irregular lines **(Fig. 1.201)**. Tape the drawing to the tablet and calibrate it as discussed previously, following these steps:

```
Command: SKETCH (ENTER)
Record increment <0.1>: 0.01 (ENTER)
Sketch. Pen eXit Quit Record Erase
Connect.
```

The record increment specifies the distances between the endpoints of the connecting lines that are sketched. Other options are:

`Pen`	raises or lowers pen.
`eXit`	records lines and exits.
`Quit`	discards temporary lines and exits.

THE SKETCH COMMAND

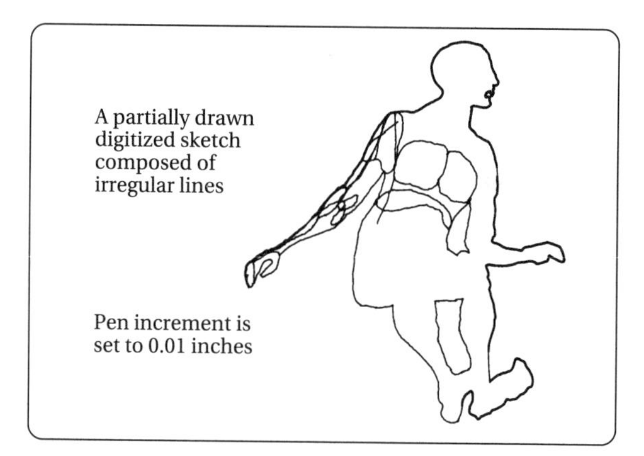

Figure 1.201 This drawing was made with the SKETCH command and a tablet instead of a mouse. The drawing was taped to the tablet and traced with the stylus with increments of 0.01 inches.

Record	records temporary lines.
Erase	deletes selected lines.
Connect	joins current line to last endpoint.
.	(period) draws a line from the current point to the last endpoint.

Begin sketching by moving your pointer to the first point, lower the pen (P), move the stylus over the line, and the line is sketched on the screen. To erase, raise the pen (P), enter ERASE (E), move the stylus backward from the current point, and select the point where the erasure is to stop. All lines are temporary until you select Record (R) or eXit (X) to record them. Begin new lines by repeating these steps.

The SKPOLY variable can be set by typing SETVAR, entering SKPOLY, and typing 1 (On). When using the SKETCH command with SKPOLY, lines will be drawn as continuous polylines—as an alternative to lines that are connected with individual line segments.

OBLIQUE PICTORIAL

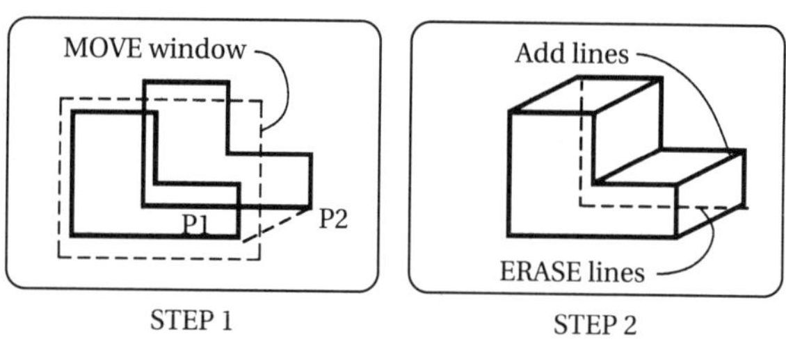

Figure 1.202 An oblique pictorial.
Step 1 Draw the frontal surface of the oblique and COPY this view from P1 to P2.
Step 2 Connect the corner points and erase the invisible lines to complete the oblique.

TYPES OF GRIDS

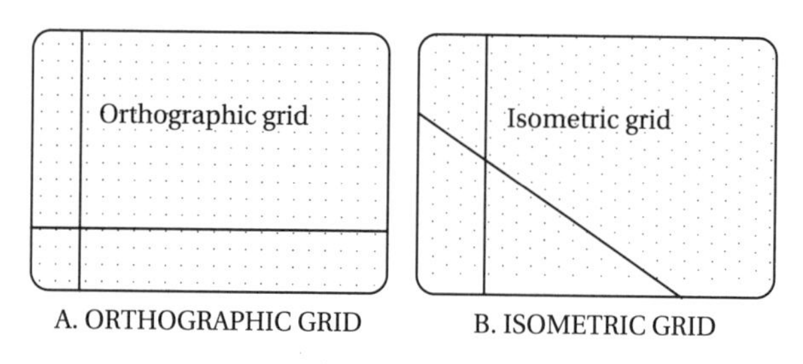

Figure 1.203
A The orthographic grid is called the Standard style of the SNAP mode.
B The Isometric style of the SNAP mode.

1.69 Oblique Pictorials

An oblique pictorial can be constructed as shown in **Fig. 1.202**. The front orthographic view is drawn and COPYed behind the first view at the angle selected for the receding axis. The visible endpoints are connected with OSNAP on and invisible lines are erased. Circles are drawn as true circles on the true-size front surface, but circular features should be avoided on the receding planes since their construction is complex.

ISOMETRIC DRAWING

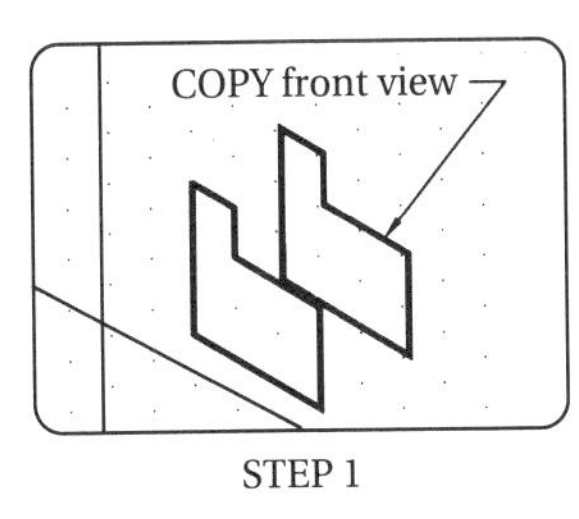

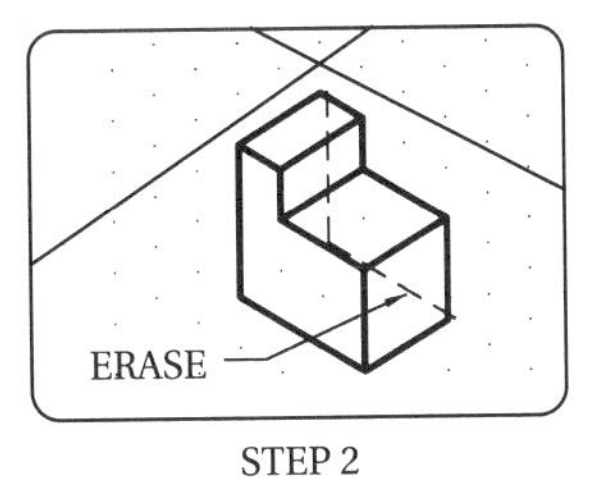

Figure 1.204 The isometric pictorial.
Step 1 Draw the front view of the isometric and `COPY` it at its proper depth.
Step 2 Connect the corner points, and erase the invisible lines. The cursor lines can be moved into three positions using (CTRL) `E` or `ISOPLANE`.

1.70 Isometric Pictorials

The `Style` option of the `SNAP` command is used to change the rectangular `Grid` from `Standard (S)` to `Isometric (I)` to show the grid dots in an isometric pattern (vertically and at 30° with the horizontal) **(Fig. 1.203)**. The cursor will align with two of the isometric axes and can be `SNAP`ped to the grid. The axes of the cursor are rotated 120° by pressing (CTRL) `E`, (F5), or by using the `ISOPLANE` command. When `ORTHO` is `On`, lines are forced parallel to the isometric axes. **Figure 1.204** shows the steps for constructing an isometric drawing.

When the `Grid` is set to the isometric mode, the `ELLIPSE` command will give the following options:

`<Axis endpoint 1>/Center/Isocircle:` <u>`I`</u> (ENTER)

Select the `Isocircle` option and ellipses are positioned in one of the isometric orientations shown in **Fig. 1.205** by using (CTRL) `E`. An example of an isometric with partial ellipses is shown in **Fig. 1.206**.

The oblique and isometric drawings covered here are 2D drawings that appear to be 3D views, but they cannot be rotated on the screen to obtain different view points.

ELLIPSES IN ISOMETRIC

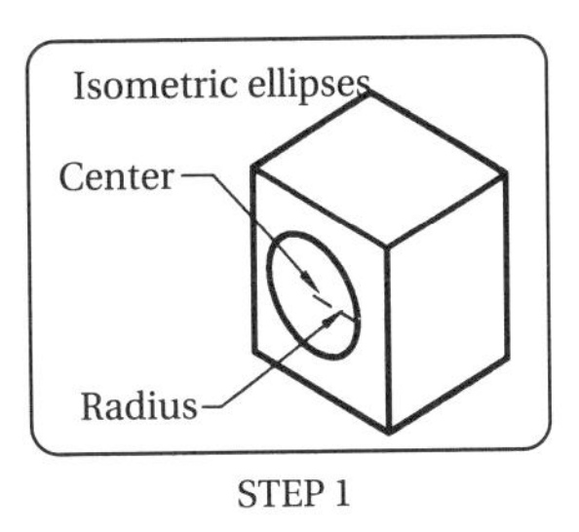

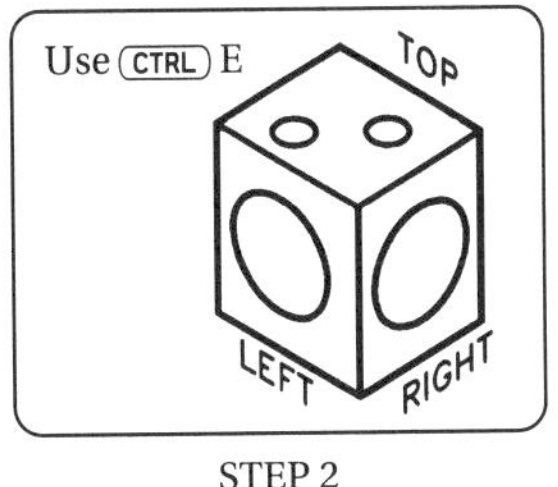

Figure 1.205 `ELLIPSE`: Isometric mode.
Step 1 When in the isometric `SNAP` mode:
`Command:` <u>`ELLIPSE`</u> (ENTER)
`<Axis endpoint 1>/Center/Isocircle:` <u>`I`</u> (ENTER)
`Center of circle:` (Select center.)
`<Circle radius>/Diameter:` (Select radius.)
Step 2 The isometric ellipse is drawn on the current `ISOPLANE`.

CIRCULAR FEATURES IN ISOMETRIC

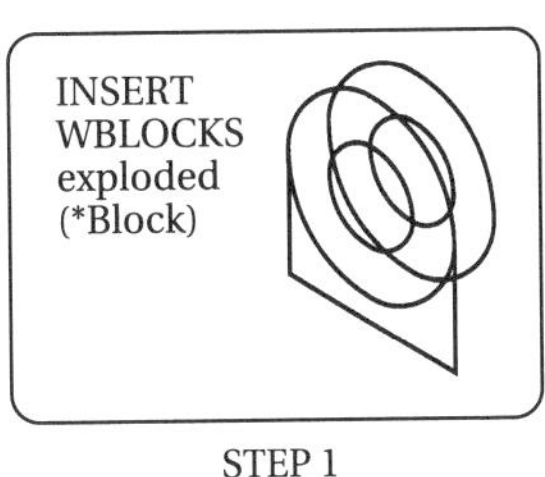

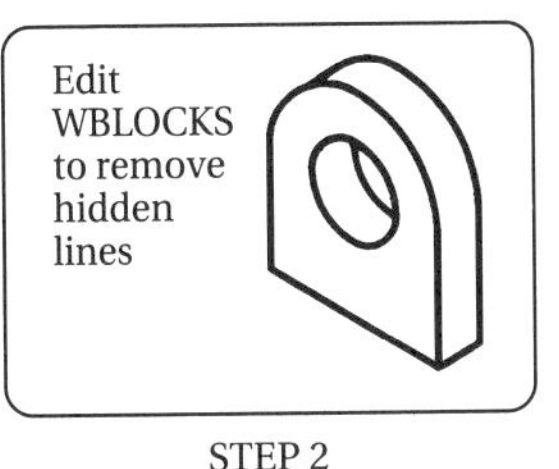

Figure 1.206 `ELLIPSE`: Isometric.
Step 1 The `Unit Block` of the ellipse developed in Fig. 1.205 is inserted and sized (X-scale factor) to match the drawing's dimensions. The ellipse can be rotated as a `Block` to fit any of the three isometric surfaces.
Step 2 The back side of the isometric is drawn. The hidden lines are removed with the `BREAK` command.

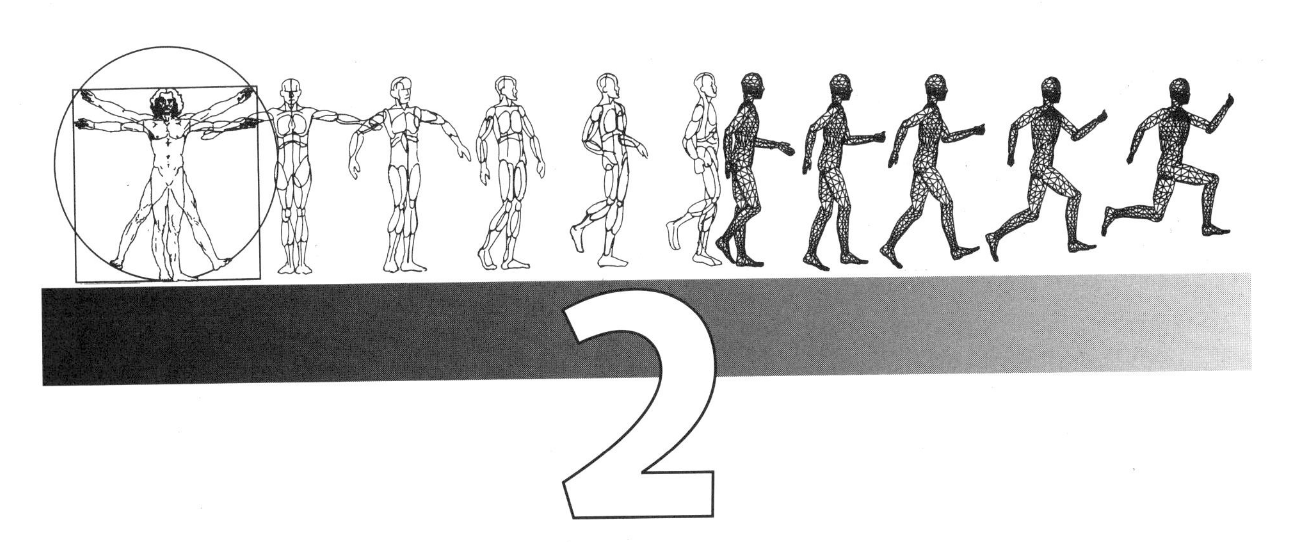

3D Drawing, Solid Modeling, and Rendering with AutoCAD Release 13

2.1 Introduction

This chapter provides an introduction to the principles of making true 3D pictorial drawings that can be rotated on the screen and viewed from any angle as if they were held in your hand. The three major divisions of this chapter are fundamentals of 3D drawing, solid modeling, and rendering.

2.2 Paper Space and Model Space: An Overview

The two distinctly different ways of obtaining views of objects are with `TILEMODE=1` (`On`) or `TILEMODE=0` (`Off`). When `TILEMODE` is on, the screen can be divided into viewports (`VPORTS`) that abut each other in standard arrangements as do flooring tiles **(Fig. 2.1A)**. When `TILEMODE` is off, the viewports are created with `MVIEW` in standard or nonstandard positions as shown in **Fig. 2.1B**.

TILEMODE ON AND OFF

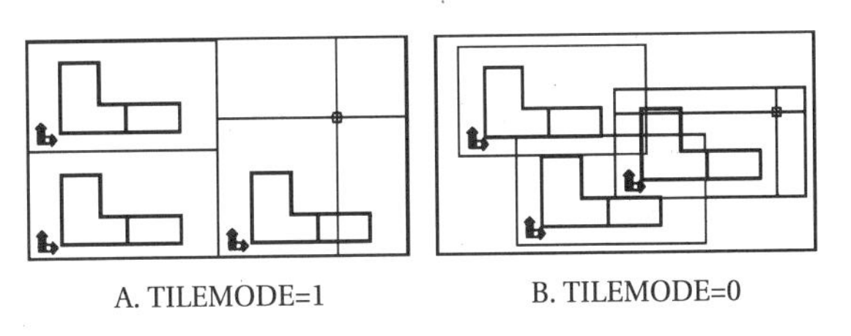

Figure 2.1
A When `TILEMODE=1` (`On`) the viewports are arranged to abut each other like flooring "tiles."
B When `TILEMODE=0` (`Off`) 3D viewports are made with `MVIEW` (make view) which can overlap.

TILEMODE On

By setting `TILEMODE=1` (`On`), the screen enters `Model Space` by default and drawings can be made in two dimensions (2D) and three dimensions (3D). The `VPORTS` command lets you create up to four tiled viewports at a time on the screen,

TILEMODE ON

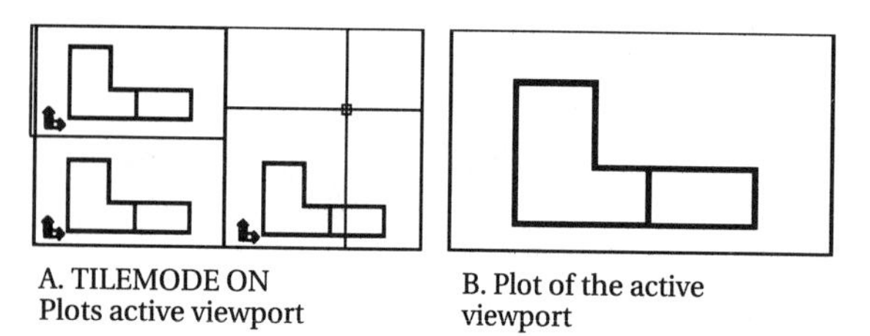

A. TILEMODE ON
Plots active viewport
B. Plot of the active
viewport

Figure 2.2 When `TILEMODE` is set to `1 (On)`, only the active viewport can be plotted to paper.

TILEMODE ON AND OFF

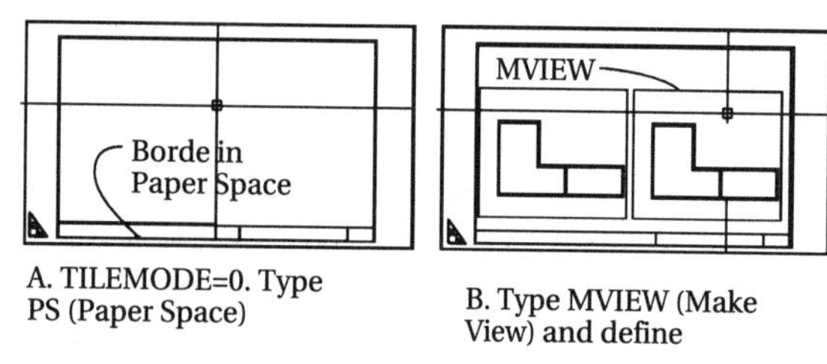

Figure 2.3

A When `TILEMODE` is set to zero and `PS` is typed, the screen is in paper space where a 2D border can be drawn.

B Use the `MVIEW` command to define the diagonal corners of 3D viewports. Create a second viewport in the same manner.

TILEMODE ON AND OFF

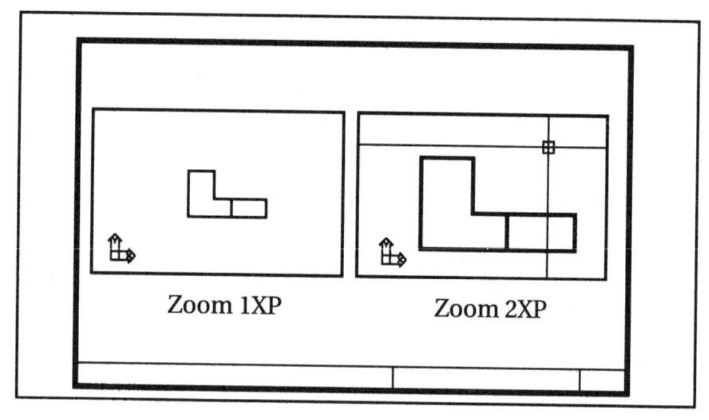

Figure 2.4 Type `MS` to enter model space and type `ZOOM` and `1XP` to show the drawing full-size in that viewport. Select the other viewport, then type `ZOOM` and `2XP` to obtain a double-size view in that viewport.

PLOTTING FROM PAPER SPACE

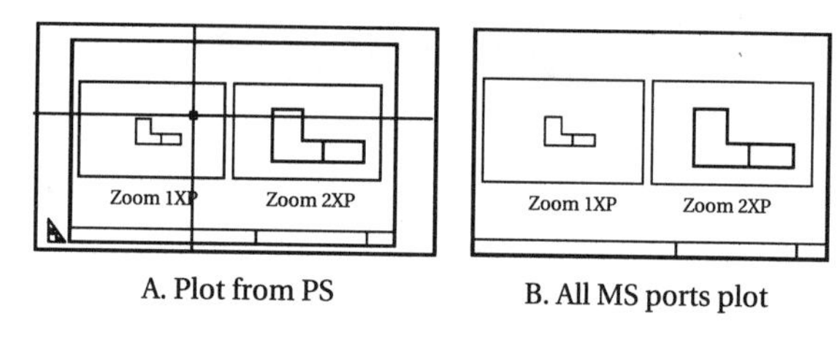

Figure 2.5

A Type `PS` to enter paper space for plotting a drawing.

B Both paper space and model space are plotted.

and more when viewports are subdivided. Drawings made before applying the `VPORTS` command will be duplicated in each viewport as if multiple monitors were wired to your computer **(Fig. 2.2)**. Only the viewport in which the cursor appears is active and can be plotted.

TILEMODE Off

Set `TILEMODE=0 (Off)`, type `PS` to enter paper space (2D space), the word `PAPER` appears in the status bar at the bottom of the screen, and a triangle icon appears at the lower left of the drawing area. Since drawings made in `Paper Space (PS)` are 2D drawings, now is a good time to insert a 2D border **(Fig. 2.3A)**.

Type `MVIEW` and create two model-space windows by selecting their diagonal corners **(Fig. 2.3B)**. Type `MS` or double click on the `Paper` button at the bottom of the screen to enter model space (`MODEL`) and the cursor will appear in the active viewport. Move to a new viewport and select it with the cursor. The drawing in the model-space port is scaled by typing `ZOOM` and `1XP` for a full-size drawing, `2XP` for a double-size drawing, and `0.5XP` for a half-size drawing **(Fig. 2.4)**.

To make a plot, return to `Paper Space` by typing `PS`; the triangular icon reappears, and `PAPER` replaces `MODEL` at the bottom of the screen. Plot from paper space and both the 2D and 3D drawings will plot as they appear on the screen **(Fig. 2.5)**.

2.3 Paper Space Versus Model Space

The following points summarize what you can do from `Paper Space` (`PS`):

1. `MVIEW` makes MS viewports.
2. `STRETCH`, `MOVE`, and `SCALE` MS viewports.
3. `ERASE` MS viewports.
4. `FREEZE` MS outlines.
5. `INSERT` 2D drawings.
6. `HIDEPLOT` removes invisible lines from selected viewports.
7. `TEXT` can be added across MS viewports.

The following points summarize what you can do from `Model Space` (`MS`):

1. Modify a 3D drawing.
2. `ROTATE` the User Coordinate System (UCS).
3. `PAN`, `ZOOM`, `SCALE`, etc., MS drawings.
4. Attach dimensions to the MS drawing.
5. `ERASE` the contents of an MS viewport.

2.4 Fundamentals of 3D Drawing

To experiment with 3D drawing, set `TILEMODE=1` (`On`), type `UCSICON`, set to `On`, and the `XY` icon appears as shown in **Fig. 2.6**. Select the `ORigin` option of the `UCSICON` command to make it appear at the origin and a plus sign appears in its corner. A `W` appears on the icon when it represents the **World Coordinate System (WCS)**. Without the `W`, you are in the **User Coordinate System (UCS)**. The **broken-pencil** icon warns that the projection plane appears as an edge, making drawing in that viewpoint impractical. The oblique-box icon indicates that the current drawing is a perspective. The triangular icon tells you the screen is in 2D paper space. `UCSICON` never shows the Z-axis; it is found by the right-hand rule shown in **Fig. 2.7**.

The various `Tiled Viewports`, found under `View` of the main menu, and `Layout` option are

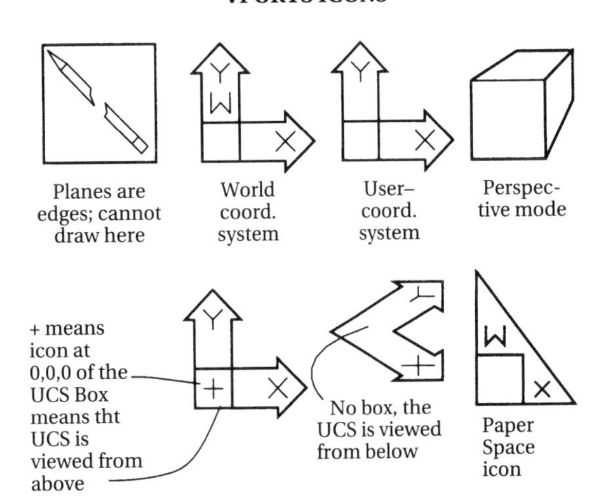

Figure 2.6 The various `VPORTS` icons that appear on the screen to show the X and Y-axes.

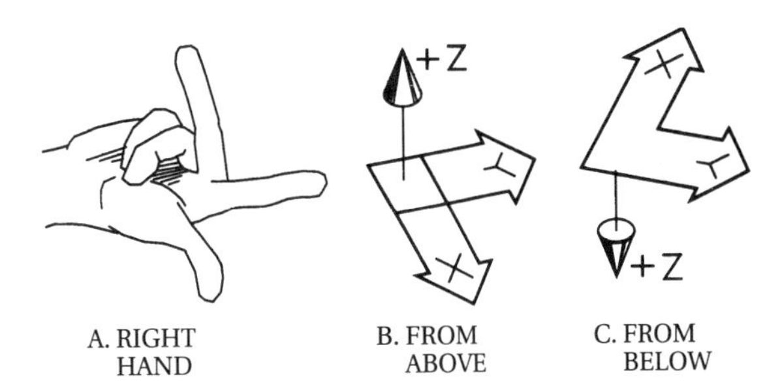

Figure 2.7 By pointing the thumb of your right hand in the positive X direction and your index finger in the positive Y direction, your middle finger points in the positive Z direction.

shown in **Fig. 2.8**. Use the `VPORTS` save option to save `VPORT` arrangements.

2.5 Elementary Extrusions

The elementary extrusion technique is a method of drawing 3D objects with the `ELEV`, `THICKNESS`, `PLAN`, and `HIDE` commands.

`ELEV` (elevation) sets the level of the base plane of the drawing.

TILED VIEWPORT LAYOUT DIALOG BOX

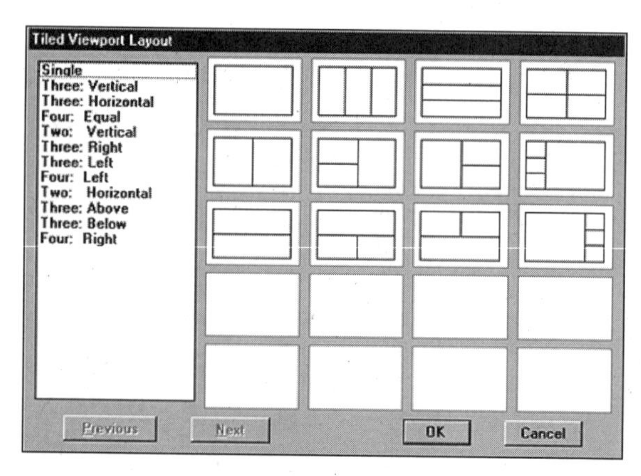

Figure 2.8 `Tiled Viewports` and its `Layout` option found under `View` on the main menu can be used for displaying viewports on the screen.

`THICKNESS` is the distance of the extrusion parallel to the Z-axis and perpendicular to the base plane.

`PLAN` changes the UCS to give a true-size view of the XY icon and all surfaces parallel to it.

`HIDE` removes invisible lines of the extruded surfaces.

Turn the `UCSICON` on the obtain the `XY` icon, type `ELEV` and set it to zero, type `THICKNESS` and set it to 4, and draw the plan view of the object with the `LINE` command **(Fig. 2.9)**. The X- and Y-axes are true size in the `PLAN` view (top view), and the Z-axis is perpendicular to them and points toward you.

Type `VPOINT` and specify a line of sight with settings of `1, -1, 1` to obtain an isometric view of the extruded block in Fig. 2.9. When you type `PLAN`, the `UCSICON` is shown true size and planes of the object that are parallel to it appear true size, also. You may use all the regular `DRAW` commands such as `LINE`, `CIRCLE`, and `ARC` to draw features, all of which will be extruded 4 units in the Z-direction. The extrusion value of 4 units will remain in effect until reset.

`HIDE` and press ENTER to remove hidden lines

EXTRUSION OF A BOX

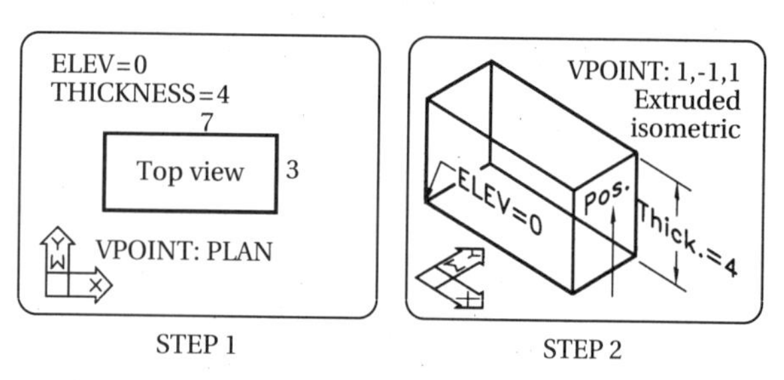

Figure 2.9 Extrusion of a box:
Step 1 `Command:` `ELEV` ENTER
`New current elevation <0>:` `0`
`New current thickness <0>:` `4`
`Command:` `LINE` ENTER
`Line From point:` (Draw 7 x 3 rectangle as a top view.)
Step 2 `Command:` `VPOINT` ENTER
`Rotate/<View point><current>:` `1, -1, 1` ENTER
(An isometric view of the extruded box appears.)

HIDING AN EXTRUDED BOX

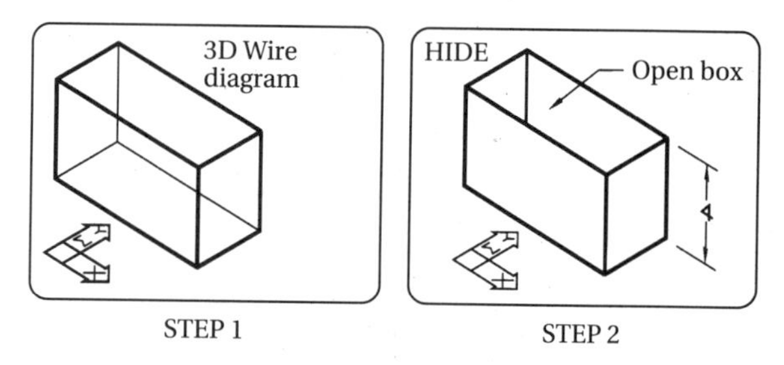

Figure 2.10 Extrusion (`HIDE` option):
Step 1 (After the box is drawn, the box appears as a wire diagram.)
Step 2 `Command:` `HIDE` ENTER
[Vertical surfaces are opaque (solid) and the top appears open.]

from an extruded object and give an "empty-box" look **(Fig. 2.10)**. Type `3DFACE` and `OSNAP` to the corner points on the upper surface to make the top of the box opaque when the `HIDE` command is applied **(Fig. 2.11)**.

The `SOLID` command can be used to make the

HIDING A 3DFACE

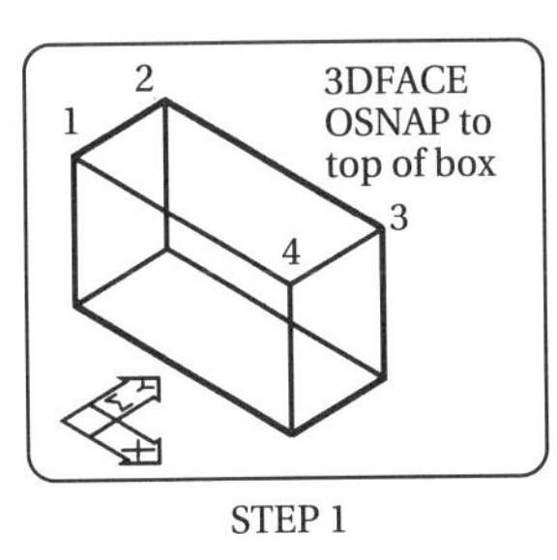

STEP 1

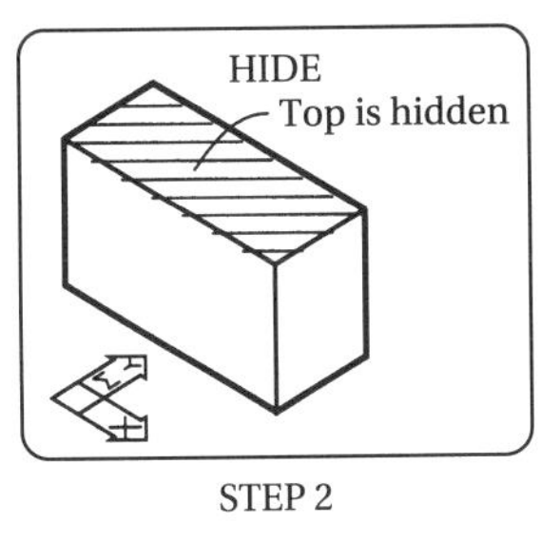

STEP 2

Figure 2.11 Extrusion (`3Dface` option):

Step 1 `Command:` `OSNAP` ENTER
`Object snap modes:` `END` ENTER
`Command: 3DFACE` ENTER
`First point:` `1`
`Second point:` `2`
`Third point:` `3`
`Fourth point:` `4`
`Third point:` ENTER

Step 2 `Command:` `HIDE` ENTER
(The top surface appears as an opaque surface.)

top surface opaque by assigning it an `ELEV` equal to the extruded `THICKNESS` (4), setting its `THICKNESS` to zero, and applying a solid area to the top by selecting the four corners. Type `HIDE` to make the top appear opaque.

2.6 Coordinate Systems

Lines are drawn in the plane of the coordinate system indicated by an `XY` icon. The two coordinate systems are the World Coordinate System (WCS) and the User Coordinate System (UCS).

World Coordinate System (WCS) has an origin where X, Y, and Z are 0 and, usually, the X- and Y-axes are true length in the top, or plan, view. The `UCSICON` has a `W` and a plus sign on it when the icon is at the `WCS` origin.

A **User Coordinate System (UCS)** can be located within the `WCS` with its X- and Y-axes positioned in any direction and its origin at any selected point.

Type `UCSICON` and `On`, select `Origin` and move

UCS: 3P OPTION

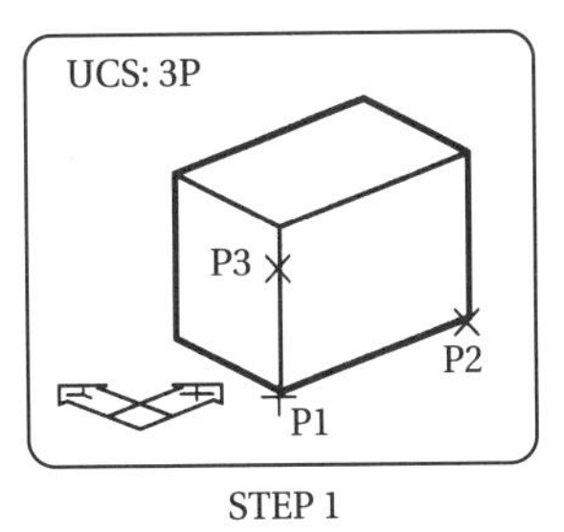

STEP 1

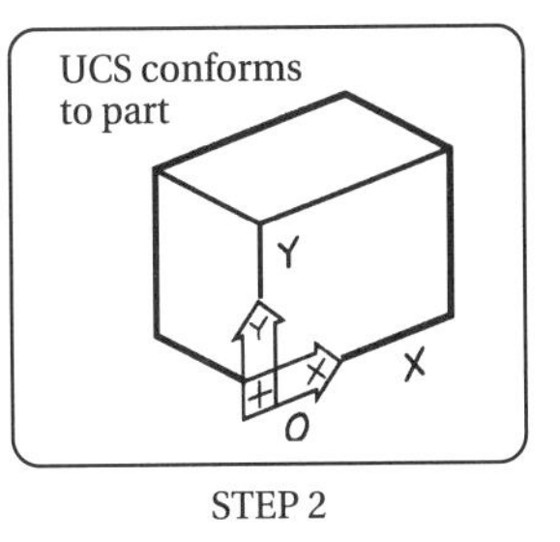

STEP 2

Figure 2.12 UCS (`3Point` option):

Step 1 `Command:` `UCS` (`3point` option) ENTER
`Origin point <0, 0, 0>:` `P1` (Origin)
`Point on positive portion of the X axis:` `P2`
`Point on positive Y portion of UCS X-Y plane:` `P3`

Step 2 The `UCS` icon is transferred to the origin. The plus sign at its corner box indicates that it is at the origin.

it to its origin, and establish a User Coordinate System in the following manner:

`Command:` `UCS` ENTER
`Origin/ZAxis/3point/OBject/`
`View/X/Y/Z/Prev/Restore/Save/`
`Del/?/<World>:` `O` ENTER
`Origin point <0,0,0>:` (Select.) ENTER

The options of the `UCS` command are:

`World`: Set the system to the World Coordinate System.

`Origin`: Define an origin without changing the orientation of the X-, Y-, and Z-axes by picking a point.

`ZAxis`: Select a new origin and pick a point on the positive portion of the new Z-axis to locate a new UCS.

`3point`: Select a new origin, a point on the X-axis, and a point on the Y-axis to establish a new `UCS` **(Fig. 2.12)**.

`Object`: Pick an object (other than `3D Polyline` or a polygon mesh) and the `UCS` will have the same positive Z-axis as the selected entity.

View: Select View to establish a UCS in which the XY-plane is parallel to the screen, which allows the application of 2D text to a 3D drawing.

X/Y/Z: Specify X, Y, or Z as the axis about which to rotate the UCS and type the angle of rotation. Example: Type X and press (ENTER), type 90 and press (ENTER), and the icon is rotated 90° about the X-axis.)

Prev (Previous): The previous UCS is recalled.

Restore: Select R and type the name of the saved UCS and it becomes the current UCS.

Save: Select S and name the current UCS to save it.

Del (Delete): Select D and give the name of the UCS to delete it.

?: Pick ? to get a listing of the current and saved coordinate systems. If unnamed, the current UCS is listed as *WORLD* or *NO NAME*.

The UCS Control dialog box, found by typing DDUCS or by selecting View and Named UCS, lists the saved coordinate systems. The *WORLD* coordinate system is listed first. The current UCS is indicated by CURRENT, but a new one can be selected with the cursor (SETUP2, for example), or by picking the Current box. Delete a UCS by picking the Delete box, or rename one by picking the Rename To box and typing a new name. OK confirms any action made and Cancel closes the dialog box. When List is picked the UCS Origin Point and Axis Vectors box appears. The box shows the coordinates of the origin and the endpoints of the three axes. Select OK to return to the UCS Control box.

The UCSICON is turned on in the following manner:

```
Command: UCSICON (ENTER)
ON/OFF/ALL/Noorigin/ORigin<On>: ON (ENTER)
```

The functions of these UCSICON options are:

On/Off turns the icon on and off.

UCS ORIENTATION DIALOG BOX

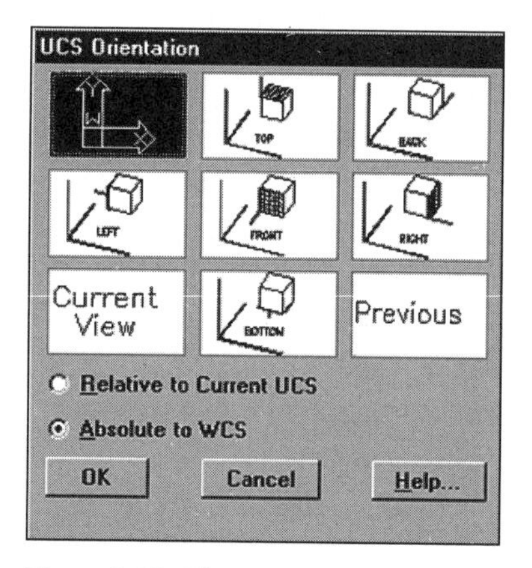

Standard views can be selected from this dialog box

Figure 2.13 The UCS Orientation dialog box is used to select standard views of objects.

All displays the icon in all viewports.

Noorigin displays the icon at the lower left corner regardless of the location of the UCS origin.

Origin places the icon at the origin of the current coordinate system if space permits, or at the lower left if space is unavailable.

2.7 Setting VPOINTS

The VPOINT command sets the viewpoint of a UCS with the UCS Orientation box, the axes, or by typing coordinates. The UCS Orientation dialog box, under View of the menu bar, is used to select UCS orientations for principal orthographic views, or the UCSICON symbol is selected for non-standard views **(Fig. 2.13)**. For example, select the top view icon, specify the origin when prompted, and type PLAN to obtain a top view of the UCSICON where the plane of the X- and Y-axes is true size.

When the UCSICON symbol is selected, the axis tripod (a set of X-, Y-, and Z-axes) appears on the screen **(Fig. 2.14)**. A viewpoint of the object is found by selecting a point on the compass globe as described in **Fig. 2.15**. Repeat this command by pressing (ENTER) and selecting other views.

VPOINT & AXES

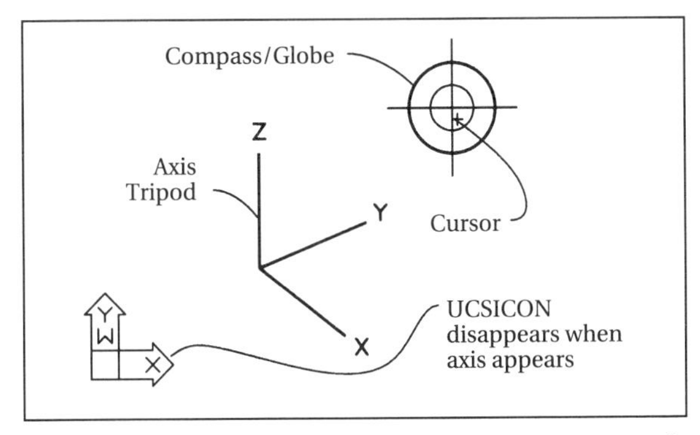

Figure 2.14 When VPOINT is typed and (ENTER) is pressed twice, a globe and a set of axes will appear on the screen for selecting a viewpoint.

THE COMPASS GLOBE

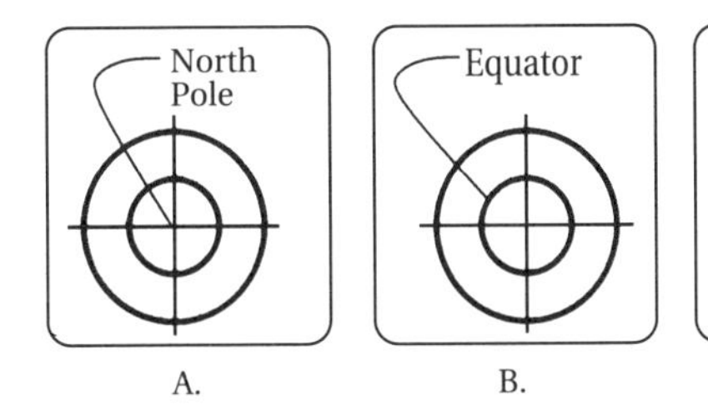

Figure 2.15

A The north pole is at the crossing of the crosshairs.

B The small circle locates viewpoint at the equator.

C The large circle locates the viewpoint at the south pole.

Figure 2.16 compares the viewpoints found on the compass globe with those specified with numbers at the keyboard. A VPOINT of 1, -1, 1 means that the origin (0,0,0) is viewed from a point that is 1 unit in the X-direction, 1 unit in the negative Y-direction, and 1 unit in the negative Y-direction, and 1 unit in the positive Z-direction from 0,0,0 **(Fig. 2.17)**.

A dialog box found by picking VIEW (menu bar)/3D Viewpoint/Rotate can be used to select a viewpoint **(Fig. 2.18)**. If you pick VIEW/3D View-point Presets, a list of viewpoint options is given.

THE COMPASS GLOBE

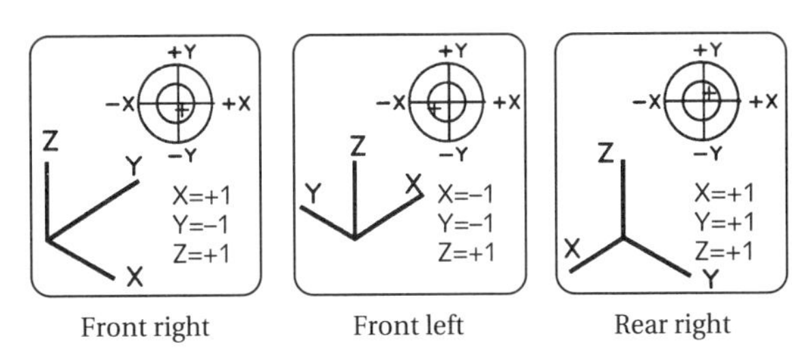

Figure 2.16 The relationships between the points on the VPOINT globe and the VPOINT values selected from the keyboard.

VPOINT POSITIONS

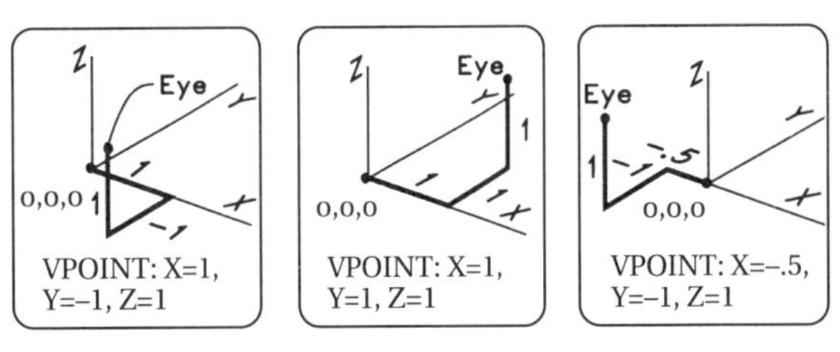

Figure 2.17 Graphical examples of the selection of VPOINTS from the keyboard.

VIEWPOINT PRESETS DIALOG BOX

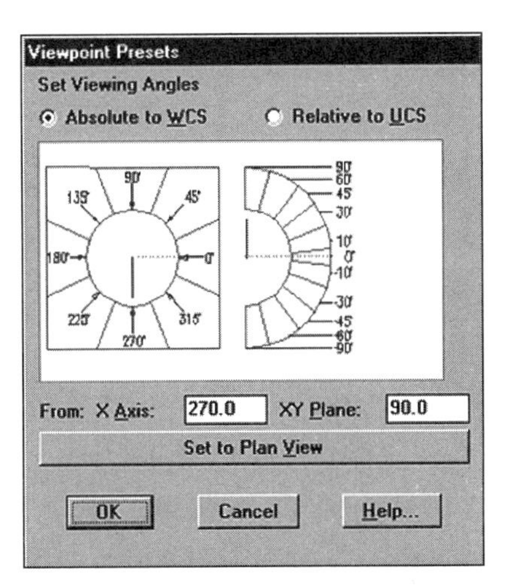

Figure 2.18 The Viewpoint Presets dialog box is used to select the point of view in the XY-plane and with the XY-plane by selecting intervals with the cursor or by typing.

MOVING THE UCS

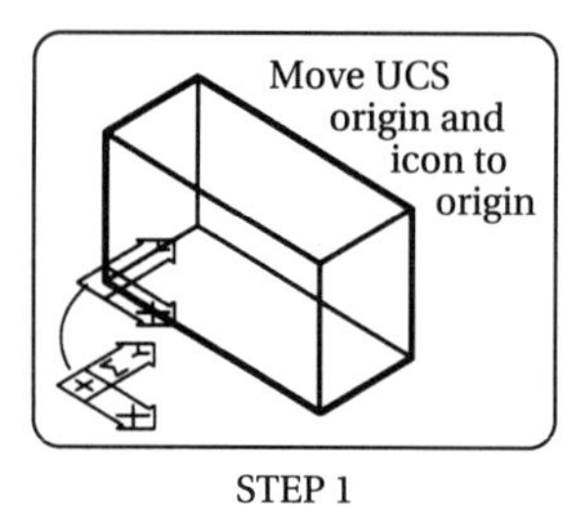

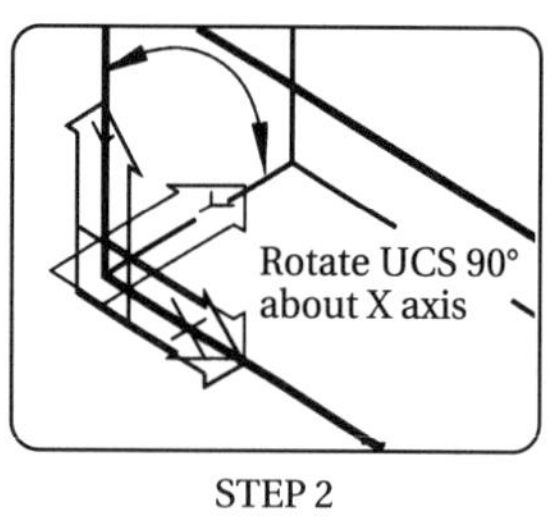

STEP 1 STEP 2

Figure 2.19 Setting the UCS:

Step 1 `Command:` <u>`UCSICON`</u> (ENTER)
`ON/OFF/All/Noorigin/ORigin<Off>:` <u>`OR`</u> (ENTER)
`Command:` <u>`UCS`</u> (ENTER)
`Origin/ZAxis/3point/Object/View/X/Y/Z/Prev/Restore/Save/Del/?/<World>:` <u>`O`</u> (ENTER)
`Origin point:` (`OSNAP` to `END` at corner of box.)

Step 2 `Command:` <u>`UCS`</u> (ENTER)
`Origin/ZAxis/3point/Object/View/X/Y/Z/...<World>:` <u>`X`</u> (ENTER)
`Rotation angle about X axis <0>:` <u>`90`</u> (ENTER)
(The icon is placed in the frontal plane of the box.)

The six principal orthographic views can be picked from `3D Viewport Presets` under `View`, or by typing from the `VPOINT` command as follows:

Presets	By Typing
Top view	`0,0,1`
Front view	`0,-1,0`
Right-side view	`1,0,0`
Left-side view	`-1,0,0`
Back view	`0,1,0`
Bottom view	`0,0,-1`

2.8 Application of Extrusions

An extruded box, similar to the one in Fig. 2.9, is shown in isometric by typing `VPOINT`, pressing (ENTER), and giving coordinates of `1,-1,1` in **Fig. 2.19**. The `UCSICON` is moved to the object's lower

MOVING THE UCS

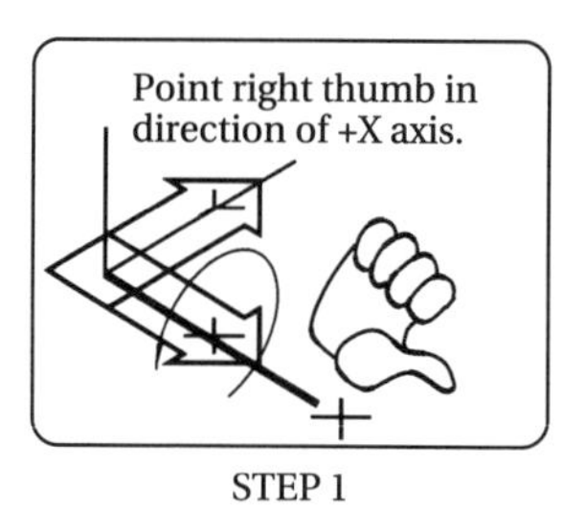

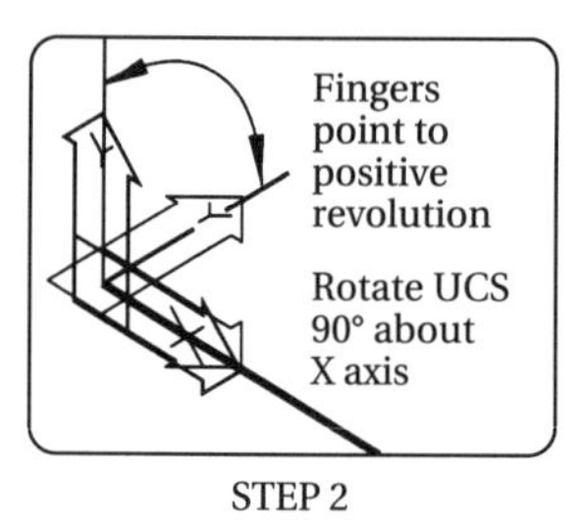

STEP 1 STEP 2

Figure 2.20 Rotating the UCS:

Step 1 `Command:` <u>`UCS`</u> (ENTER)
`Origin/ ZAxis.../<World>:` <u>`X`</u> (ENTER)

Step 2 `Rotation angle about X axis <0>:` <u>`90`</u> (ENTER)
(The UCS rotates 90° about the X-axis.)

left corner by the `UCS` command and rotated 90° about the X-axis to lie in the frontal plane of the box. The right-hand rule is used to determine the direction of rotation by pointing your right thumb in the positive direction of the axis of rotation **(Fig. 2.20)**. Type `PLAN`, and the `UCSICON` and front view will appear true size.

The circle is drawn as an extrusion by setting the `ELEV` to `0` and `THICKNESS` to `-3` (the depth of the box) and drawing a cylindrical hole **(Fig. 2.21)**. Apply `3DFACEs` to the upper and lower planes of the box and `HIDE` invisible lines.

2.9 DVIEW: Dynamic View

The `DVIEW` command is similar to the `VPOINT` command, but with `DVIEW` the viewpoints of an object are seen as they are dynamically changed by the cursor. In addition to axonometric views (parallel projections), three-point perspectives can be obtained for the most realistic pictorials **(Fig. 2.22)**. The `DVIEW` command has the following options:

```
Command: DVIEW (ENTER)

CAmera/TArget/Distance/POints/PAn/Zoom/
TWist/CLip/Hide/Off/Undo/<eXit>:
```

EXTRUDING A HOLE

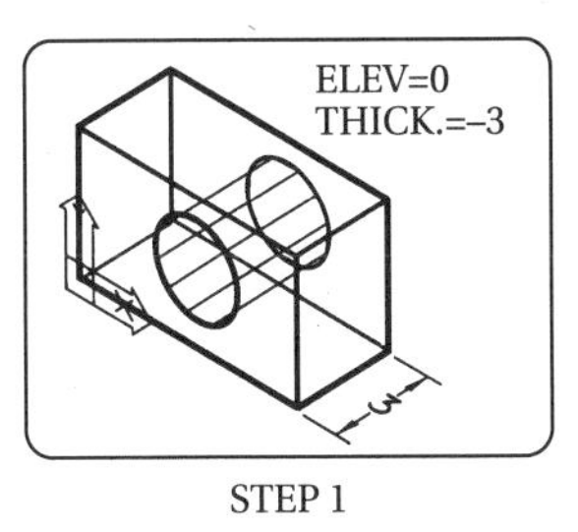

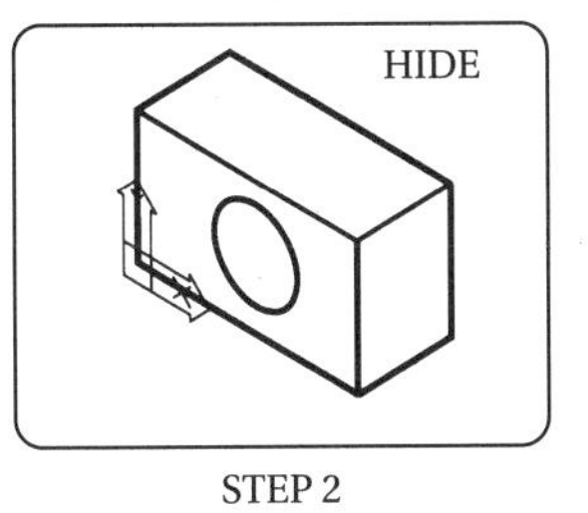

Figure 2.21 Extruding a hole:

Step 1 `Command:` `ELEV` (ENTER)
`New current elevation <0>:` `0` (ENTER)
`New current thickness <4>:` `-3` (ENTER)
`Command:` `CIRCLE` `3P/2P/TTR/<Center point>:` (ENTER)
(Select center with cursor.)
`Diameter/<Radius>:` (ENTER) `.5` (A 1" diameter cylinder is extruded 3" deep into the box.)

Step 2 `Command:` `HIDE` (ENTER) (The outline of the hole is shown, but it cannot be seen through.)

THE DVIEWBLOCK HOUSE

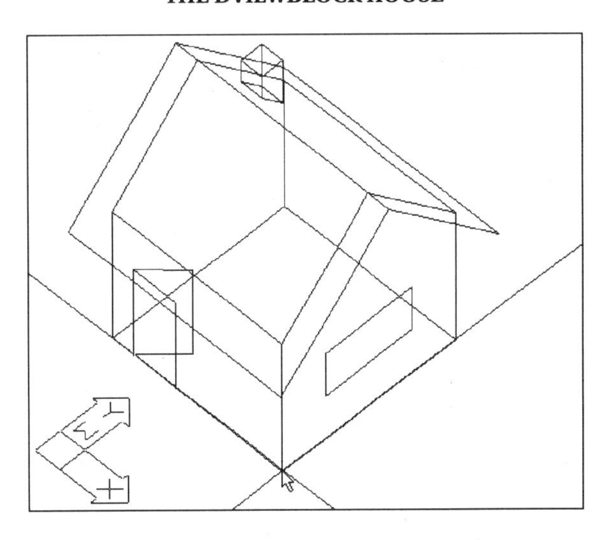

Figure 2.23 The top view of the `DVIEWBLOCK` house is obtained by typing `DVIEW` and pressing (ENTER) twice.

DISTANCE OPTION: DVIEW

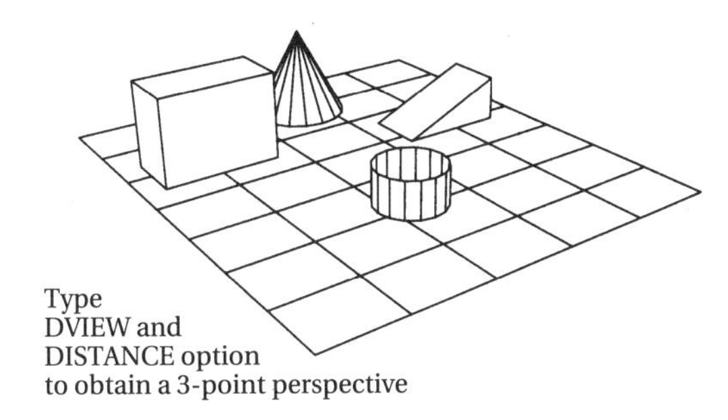

Type
DVIEW and
DISTANCE option
to obtain a 3-point perspective

Figure 2.22 A perspective view is obtained when the `Distance` option of the `DVIEW` command is selected.

CAMERA OPTION: DVIEW

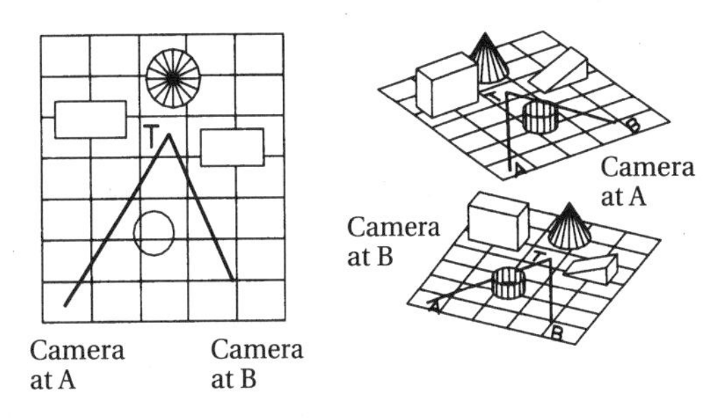

Figure 2.24 The `CAmera` option of the `DVIEW` command is used to obtain different views of a stationary target point by moving the position of the camera.

By typing `DVIEW` and pressing (ENTER), the top view of the `DVIEWBLOCK` house appears, which can be used for experimentation with the following options **Fig. 2.23**):

`CAmera` rotates your viewpoint as if you were using a camera and were moving about the target **(Fig. 2.24)**. As the cursor (the camera) is moved, the view is dynamically changed until a viewpoint is selected.

`TArget` is identical to the `CAmera` option, but the camera remains stationary as the target (and the drawing containing it) is rotated about the camera **(Fig. 2.25)**.

TARGET OPTION: DVIEW

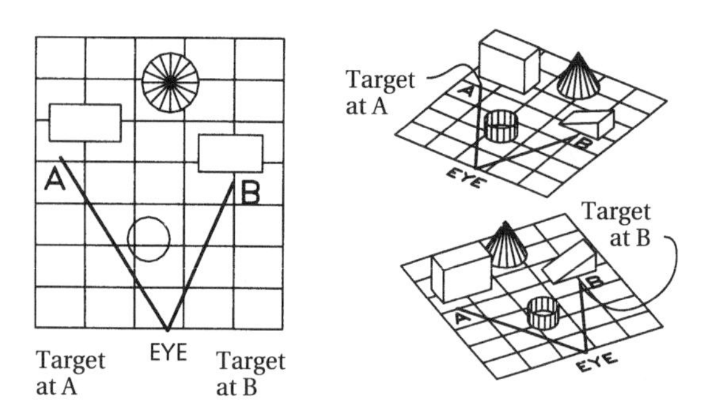

Figure 2.25 The TArget option of the DVIEW command is used to obtain different views of a scene by moving the target to different locations about a stationary camera.

CLIP OPTION: DVIEW

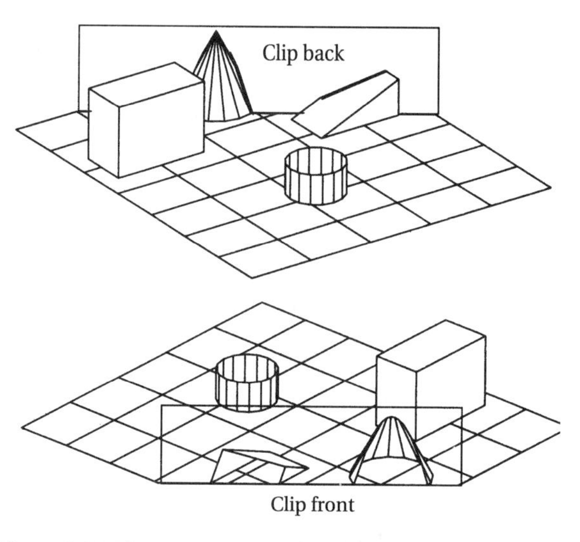

Figure 2.26 The CLip option of the DVIEW command removes Back or Front portions of a drawing. The clipping plane is parallel to the drawing screen.

Distance uses the current camera position and turns the view into a perspective. The XY icon is replaced with the perspective-box icon. When prompted, give a distance to the target by typing the value, or by using the slider bar, with a range from 0X to 16X; 1X is the current distance to the target.

POints specifies the target point and the camera position for viewing a drawing. This command is necessary to specify viewpoints for perspectives.

PAn moves the view of a drawing without changing its magnification or true position.

Zoom changes the magnification of a drawing in the same manner as the Zoom/Center command when perspective is Off. When in the perspective mode (Dist=On) is on, Zoom dynamically varies the magnification.

TWist rotates the drawing about an axis that is perpendicular to the screen.

CLip places cutting planes perpendicular to the line of sight to remove portions of a drawing either in front or back of the plane by selecting Back/Front/<Off> **(Fig. 2.26)**. Off exits from CLip. When Distance is On (perspective mode), the frontal clipping plane remains On at the camera position.

Hide suppresses the invisible lines.

OFF turns off the perspective mode enabled by Distance=On.

Undo reverses the previous DVIEW operations one at a time.

eXit ends the DVIEW command and displays the drawing.

2.10 Basic 3D Shapes

From the Surfaces toolbar **(Fig. 2.27)**, or by typing 3D, you can select a basic 3D shape or surface from the following: box, wedge, pyramid, cone, sphere, dome, dish, torus, 3D face, edge, 3D mesh, revolved surfaces, extruded surface, ruled surface, and edge surface. These shapes are wire

SURFACES TOOLBAR: 3D SHAPES

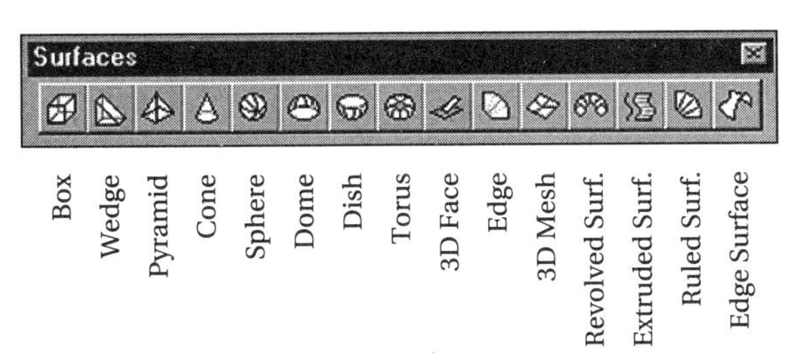

Figure 2.27 The Surfaces toolbar for drawing 3D shapes is shown here.

SURFACES: 3D BOX COMMAND

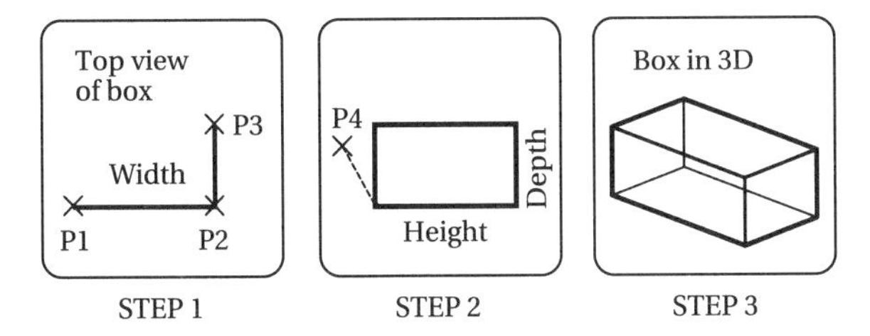

Figure 2.28 Box command:

Step 1 Command: AI_BOX (ENTER)
Corner of box: P1
Length: P2
Cube/<Width>: P3

Step 2 Height: P4

Step 3 Rotation angle about Z axis: 0 (ENTER)

(Select a VPOINT of 1,-1,1 for an isometric of the 3D box.)

frames until HIDE is used to make them appear as solids. They can be thought of as hollow shapes with meshes or faces applied to their surfaces.

BOX draws a cube or a box when you select a corner, specify length, width, and height and give the angle of rotation about the Z-axis **(Fig. 2.28)**.

WEDGE draws a wedge with the same steps used to draw the box **(Fig. 2.29)**.

PYRAMID draws a pyramid extending to its apex **(Fig. 2.30)**, or a truncated pyramid.

SURFACES: 3D WEDGE COMMAND

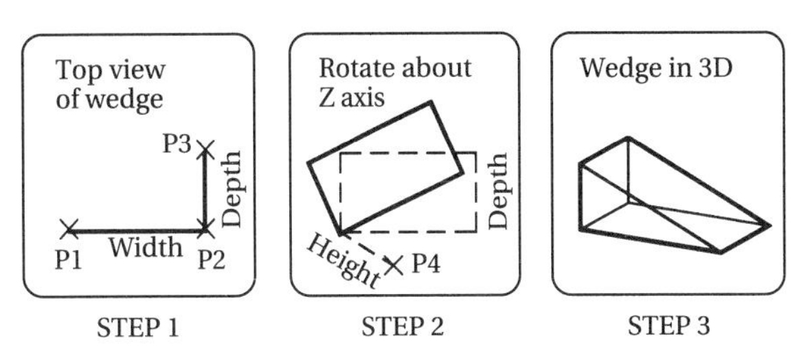

Figure 2.29 WEDGE command:

Step 1 Command: AI_WEDGE (ENTER)
Corner of wedge: P1
Length: P2
Width: P3

Step 2 Height: P4

Step 3 Rotation angle about Z axis: -15 (ENTER)

(Type VPOINT of 1,-1,1 to get the wedge in 3D.)

SURFACES: 3D PYRAMID COMMAND

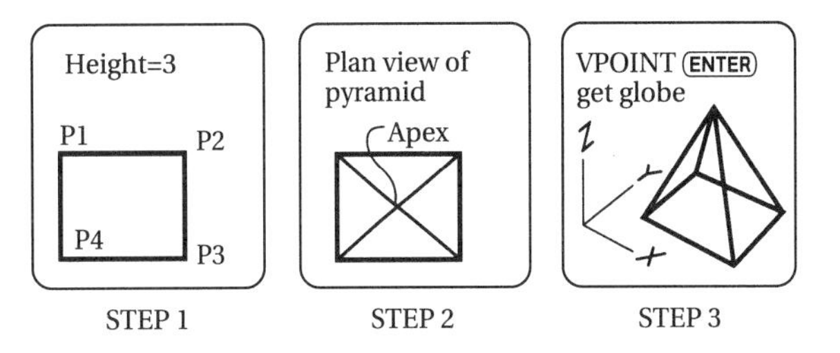

Figure 2.30 PYRAMID command:

Step 1 Command: AI_PYRAMID (ENTER)
First base point: P1
Second base point: P2
Third base point: P3
Tetrahedron/<Fourth base point>: P4

Step 2 Ridge/Top. <Apex point>: .XY of (Need Z): 2 (ENTER)

Step 3 Set the VPOINT to 1,-1,1 to get a 3D view of the pyramid.

CONE draws a cone to its apex **(Fig. 2.31)** or a truncated cone.

SPHERE draws a ball by selecting its center and radius **(Fig. 2.32)**.

SURFACES: 3D CONE COMMAND

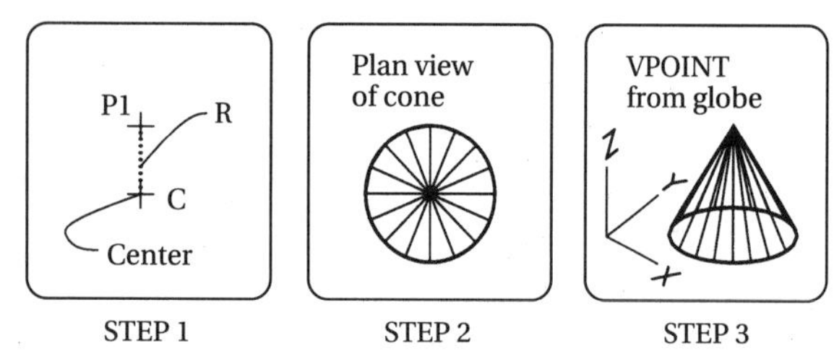

Figure 2.31 CONE command:

Step 1 Command: AI_CONE ENTER

Base center point: C

Diameter/<radius> of base: P1

Diameter/<radius> of top <0>: ENTER

Step 2 Height: 3 ENTER

Number of segments <16>: ENTER

(The plan view of the cone is generated.)

SURFACES: 3D SPHERE COMMAND

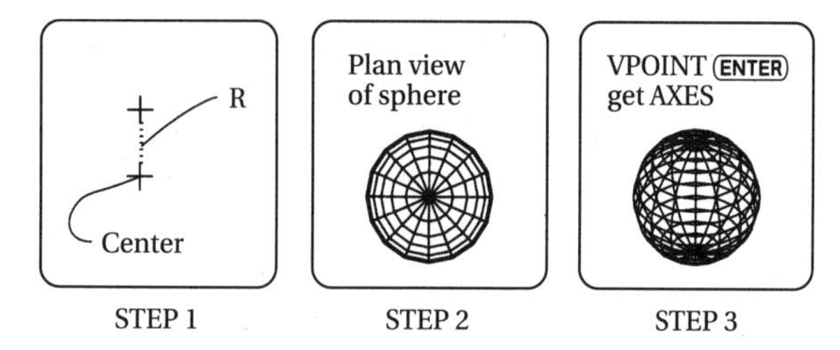

Figure 2.32 SPHERE command:

Step 1 Command: AI_SPHERE ENTER

Center of sphere: C

Diameter/<radius>: (Select with pointer.)

Number of longitudinal segments<16>: ENTER

Number of latitudinal segments<16>: ENTER

Step 2 The plan view of the sphere is generated.

Step 3 Set VPOINT to 1,-1,1 to obtain an isometric of the sphere.

DISH draws the lower hemisphere of a sphere by selecting its center and radius.

DOME draws the upper hemisphere of a sphere by selecting its center and radius in the same manner as DISH.

SURFACES: 3D TORUS COMMAND

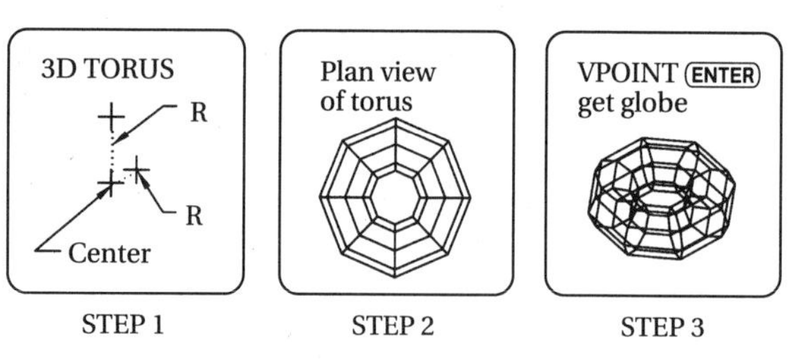

Figure 2.33 TORUS command:

Step 1 Command: AI_TORUS ENTER

Center of torus: (Select with pointer.)

Diameter/<radius> of torus: (Select with pointer.)

Diameter/<radius> of tube: (Select with pointer.)

Segments around tube circumference <16>: 8 ENTER

Segments around torus circumference <16>: 8 ENTER

Step 2 The plan view of the torus is generated.

Step 3 Set a VPOINT of 1,-1,1 for a 3D view of the torus.

TORUS draws a donut shape called a **torus** or **toroid** as shown in **Fig. 2.33**.

Other operations available from the Surfaces toolbar are covered in Section 2.68.

2.11 Surface Modeling

Faces or meshes can be applied to cover wire frames with "skins" to make them look solid. Commands from the Surfaces toolbar for applying these meshes are RULESURF, TABSURF, REVSURF, EDGESURF, and 3DMESH.

The application of these commands is illustrated by applying them to a 3D wire diagram beginning with **Fig. 2.34**. Set VPOINT to 1,-1,1 to obtain an isometric view of the 3D frame. Type SETVAR, press ENTER, type Surftab1, and set it to 20, the number of faces to be applied.

RULESURF (ruled surface) applies a surface (20 faces, as specified by Surftab1) between the circular ends (Fig. 2.34). The applied surface is MOVEd to 0,5,0 from the wire frame. RULESURF can be used to place surfaces between two objects

3D: RULESURF COMMAND

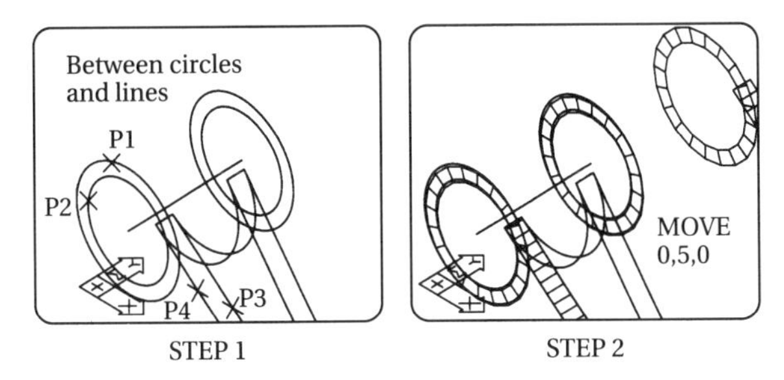

Figure 2.34 RULESURF command:

Step 1 Command: RULESURF ENTER
Select object?: P1
Select object?: P2 (Circles are faced.)
Select object?: P3
Select object?: P4 (Ends are faced.) ENTER

Step 2 Command: MOVE ENTER
Select objects? (Select faces.) ENTER
Base point or displacement: 0,5,0 ENTER
(Faces are moved.)

3D: RULESURF COMMAND

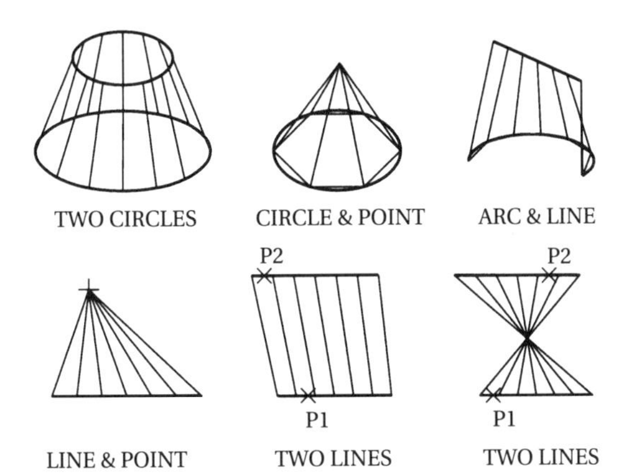

Figure 2.35 The RULESURF command connects entities with a series of 3DFACES as shown in these examples.

(curves, arcs, polylines, lines, or points) as shown in **Fig. 2.35**.

TABSURF (tabulated surface) applies a surface from a curve that is parallel and equal in length to a directional **vector (Fig. 2.36)**. Select Path Curve (the circle) and Select Direction Vector and the cylinder is drawn. The Surftab1 system variable (set to 20) applies 20 faces.

3D: TABSURF COMMAND

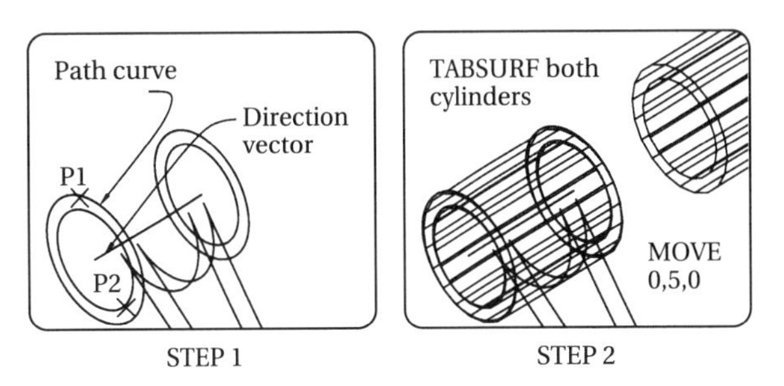

Figure 2.36 TABSURF command:

Step 1 Command: TABSURF ENTER
Select path curve: P1
Select direction vector: P2
(The spacing between the tabulated vectors is determined by the Surftab1 variable.)

Step 2 Command: MOVE ENTER
Select objects? (Select faces.) ENTER
Base point or displacement: 0,5,0 ENTER
(Faces are moved.)

REVSURF (revolved surface) revolves a line about an axis **(Fig. 2.37)** and it is MOVEd to 0,5,0 to add it to the hollow shell. **Lines**, **polylines**, **arcs**, or **circles** can be revolved to form a surface of revolution controlled by Surftab1. If a circle or a closed polyline is to be revolved (to make a torus, for example), system variable Surftab2 controls the mesh density of the line, circle, arc, or polyline that is being revolved, and Surftab1 controls the density of the path of revolution.

EDGESURF (edge surface) applies a mesh to four edges (joined boundary lines) as shown in **Fig. 2.38**. System variables Surftab1 and Surftab2 control the density of the first and second objects selected, respectively. The surfaces are moved to 0, -5, 0 to complete the hollow shell of the connector.

The beginning frame and two final views of the meshed connector are shown in **Fig. 2.39** after the HIDE command has been applied.

3D: REVSURF COMMAND

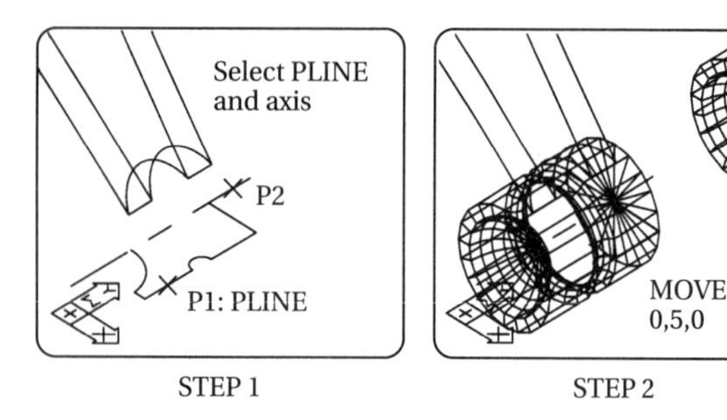

Figure 2.37 `REVSURF` command:

Step 1 `Command:` `REVSURF` ENTER
`Select path of curve:` `P1`
`Select axis of revolution:` `P2`
`Start angle <0>:` ENTER
`Included angle (+=ccw,-=cw)<Full circle>:` `360` ENTER

Step 2 `Command:` `MOVE` ENTER
`Select objects?` (Select faces.) ENTER
`Base point or displacement:` `0,5,0` ENTER
(Faces are moved.)

3D: EDGESURF COMMAND

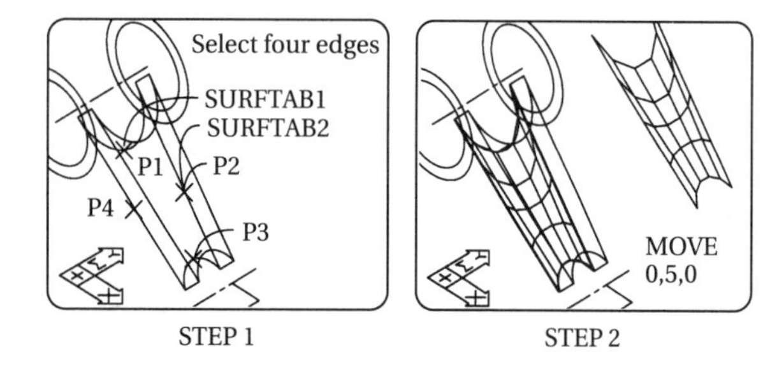

Figure 2.38 `EDGESURF` command:

Step 1 `Command:` `EDGESURF` ENTER
`Select edge 1:` `P1`
`Select edge 2:` `P2`
`Select edge 3:` `P3`
`Select edge 4:` `P4` (Mesh is drawn in the boundary. System variables `Surftab1` and `Surftab2` determine the density of the mesh.)

Step 2 `Command:` `MOVE` ENTER
`Select objects?` (Select faces.) ENTER
`Base point or displacement:` `0,5,0` ENTER
(Faces are moved.)

COMPLETED CONNECTOR 3D DRAWING

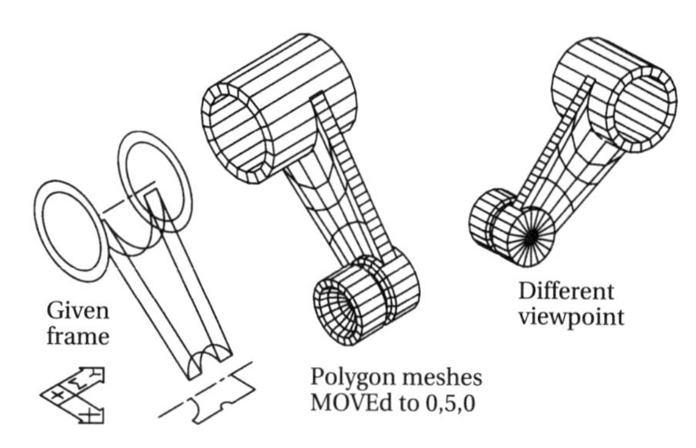

Figure 2.39 The given frame of the connector and the resulting 3D hollow shells from two viewpoints are shown here after using the `HIDE` command.

2.12 LINE, PLINE, and 3DPOLY

The `LINE` command draws 2D lines when X- and Y-values are typed, and 3D lines when X-, Y-, and Z-coordinates are given. `PLINE` draws a 2D polyline and `3DPOLY` draws 3D polylines with X-, Y-, and Z-coordinates.

`3DPOLY` or `LINE` commands draw lines with absolute coordinates from `0,0,0` located in 3D space in **Fig. 2.40**. Relative coordinates can be typed in the form of `@X,Y,Z` or `@3,0,0` to locate their endpoints with respect to the last point.

All `OSNAP` modes apply to `LINE`s, `PLINE`s, and `3DFACE`s. 3D objects can be `STRETCH`ed in the plane of the `UCSICON` (in the X- and Y-directions), but not in the Z-direction. Height dimensions are modified by rotating the UCS so height lies in the X- or Y-directions where `STRETCH` can be used.

2.13 3DFACE

`3DFACE` is used to apply faces that can be made opaque by the `HIDE` command **(Fig. 2.41)**. `3DFACE` is applied to wire frames by `SNAP`ping to their endpoints with `OSNAP`.

Figure 2.42 illustrates how corners of a `3DFACE` form an opaque plane. After selecting four points

3DPOLY: ABSOLUTE COORDINATES

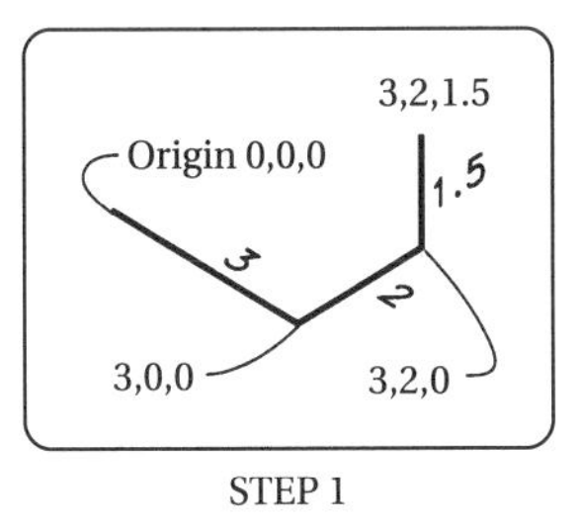

STEP 1

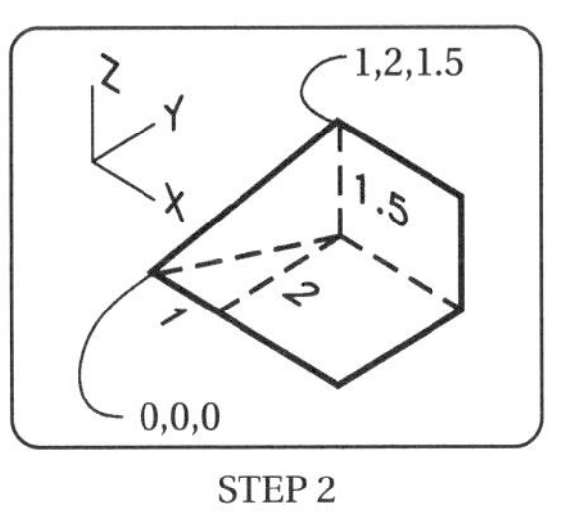

STEP 2

Figure 2.40 `3DPOLY` (absolute coordinates):

Step 1 `Command:` `3DPOLY` `(or LINE)` ENTER
`From point:` `0,0,0` ENTER
`Close/Undo/<Endpont of line>:` `3,0,0` ENTER
`/Undo/<Endpoint of line>:` `3,2,0` ENTER
`Close/Undo/<Endpoint of line>:` `3,2,1.5` ENTER

Step 2 `Close/Undo/<Endpoint of line>:` `1,2,1.5` ENTER
`Close/Undo/<Endpoint of line>:` `C` (Back to origin.) ENTER

3DFACE ON WIRE DIAGRAMS

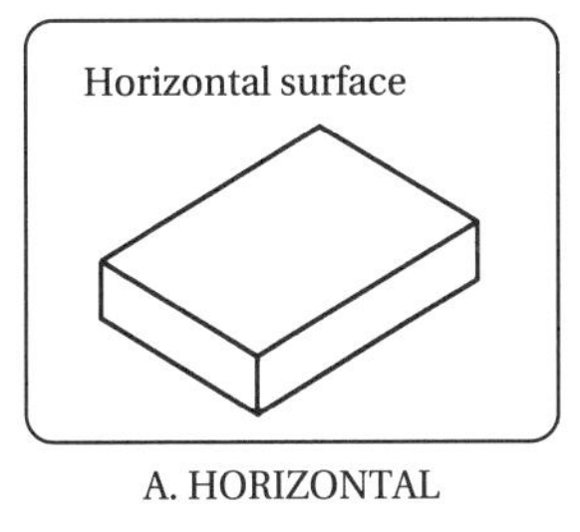

A. HORIZONTAL

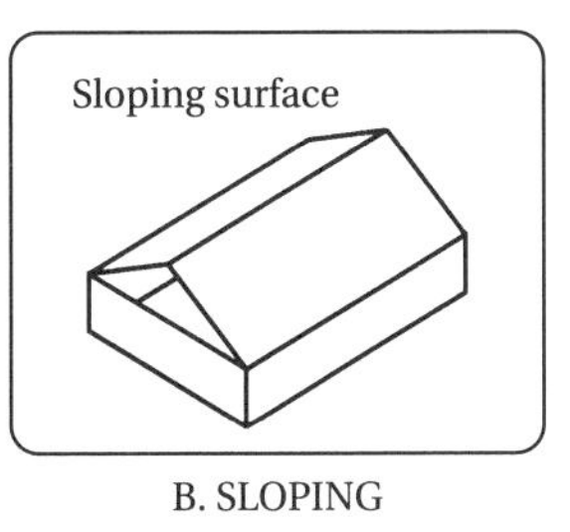

B. SLOPING

Figure 2.41 `3DFACE` is used to opaque planes of wire diagrams by snapping to endpoints and to draw opaque faces without a wire diagram. `3DFACE` is applied to a horizontal surface at **A** and to a sloping surface at **B**.

with the `3DFACE` command, you will be prompted for points 3 and 4, using the previous two points as points 1 and 2. Successively added four-sided areas are connected with splice lines yielding a "patchwork" area. By typing `I` (for invisible) and pressing ENTER prior to selecting the beginning point of a splice line, the splice will be invisible **(Fig. 2.43)**.

3DFACE COMMAND

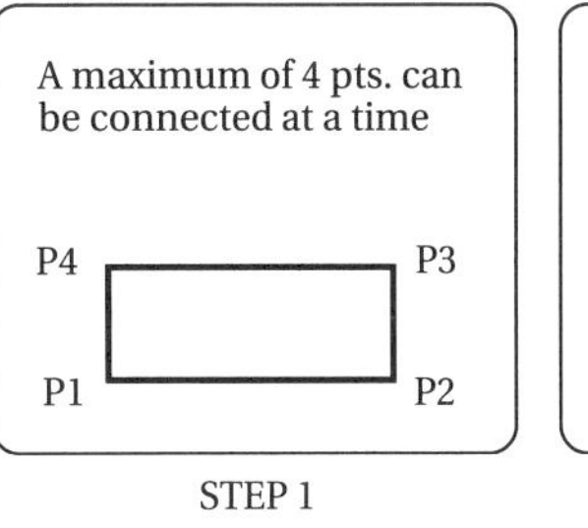

STEP 1

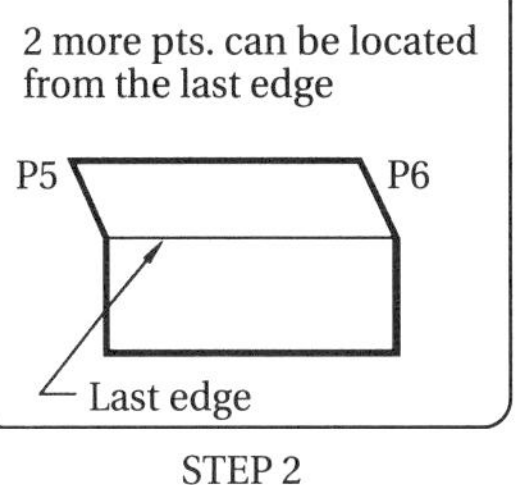

STEP 2

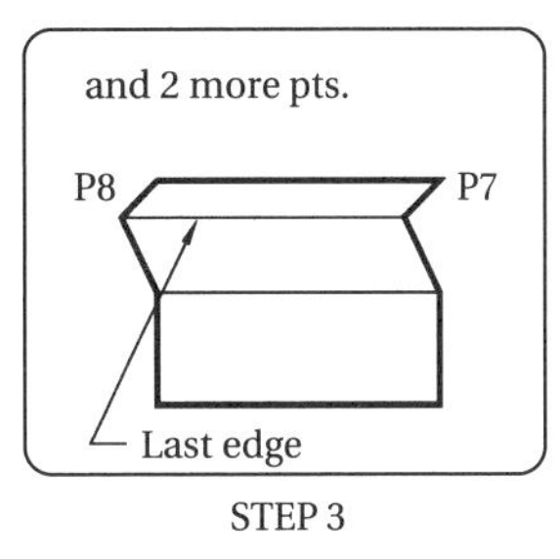

STEP 3

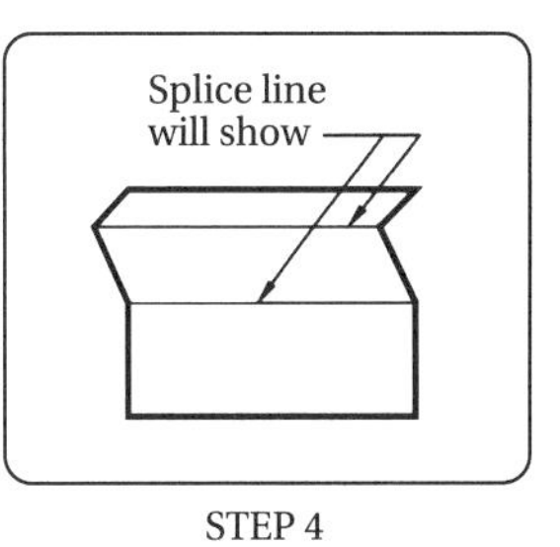

STEP 4

Figure 2.42 `3DFACE` command:

Step 1 `Command:` `3DFACE` ENTER
`First point:` `P1`
`Second point:` `P2`
`Third point:` `P3`
`Fourth point:` `P4`

Step 2 `Third point:` `P5`
`Fourth point:` `P6`

Step 3 `Third point:` `P7`
`Fourth point:` `P8`

Step 4 `Third point:` ENTER (Splice lines will show.)

2.14 XYZ Filters

Filters are used for picking points that adopt the coordinates of 3D points **(Fig. 2.44)**. When a command prompts for a point (as in the `LINE` command), type a period (`.`) followed by one- or two-letter coordinates (`.X` or `.XY` for example) and select the point with the cursor. Respond to the prompt for the missing coordinate or coordinates with a number or numbers.

To locate the front view of a point projected from the left side and top views, select the `.Y` of the

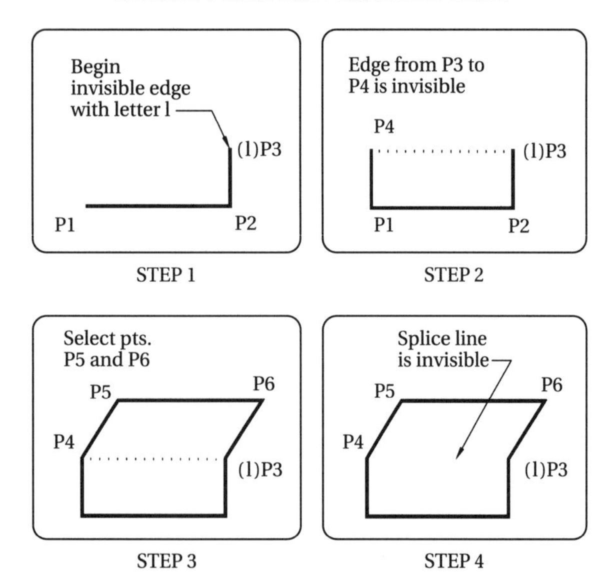

Figure 2.43 `3DFACE` (`Invisible` option):

Step 1 `Command:` <u>`3DFACE`</u> (ENTER)
`First point:` <u>`P1`</u>
`Second point:` <u>`P2`</u>
`Third point:` <u>`I`</u> (ENTER) <u>`P3`</u>

Step 2 `Fourth point:` <u>`P4`</u>

Step 3 `Third point:` <u>`P5`</u> `Fourth point:` <u>`P6`</u>

Step 4 `Third point:` (ENTER) (Splice line `P3-P4` is invisible.)

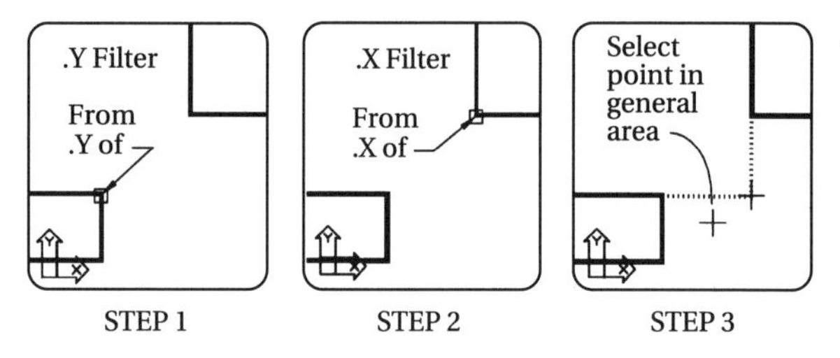

Figure 2.44 XY filters:

Step 1 `Command:` <u>`LINE`</u> (ENTER) `From point:` <u>`.Y`</u> (ENTER) `of end` (Select corner) `of` (Need XZ): <u>`.X`</u> (ENTER)

Step 2 `Close/Undo/<Endpoint of line>:` <u>`.X`</u> `of` (Select corner) `of` (Need YZ):

Step 3 Select a point in the general area of the desired position and the point appears at the intersection of the Y-coordinate from the side view and the X-coordinate from the top view.

left point **(Step 1)** and the `.X` of the top point **(Step 2)**, and the front view of the point sharing these coordinates is found **(Step 3)**.

A 3D drawing is made in **Fig. 2.45** using filters where the `.XY` coordinates of a given point are selected and `Z` is specified as `2`. `OSNAP` was set to `END` for selecting endpoints. When the points are filtered in the `XZ` plane, a prompt will ask for the Y-coordinate to locate the point in 3D space.

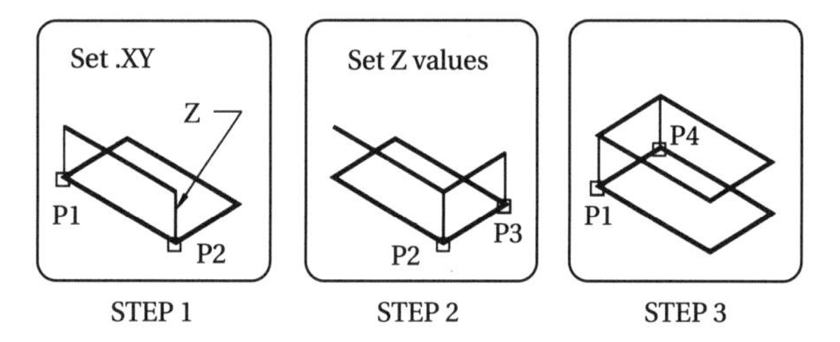

Figure 2.45 3D filters:

Step 1 Command: <u>`3DPOLY`</u> `(or LINE)` (ENTER)
`From point:` <u>`.XY`</u> `of` <u>`P1`</u> `of` (Need Z) <u>2</u> (ENTER)
`Close/Undo/<Endpoint of line>: .XY of` <u>`P2`</u> (Need Z) <u>2</u> (ENTER)

Step 2 `Close/Undo/<Endpoint of line>:` <u>`.XY`</u> `of` <u>`P3`</u> (Need Z) <u>2</u> (ENTER)

Step 3 `Close/Undo/<Endpoint of line>:` <u>`.XY`</u> `of` <u>`P4`</u> (Need Z) <u>2</u> (ENTER)
`Close/Undo/<Endpoint of line>:` <u>`.XY`</u> `of` (Select 1.) (Need Z) <u>2</u> (ENTER)

2.15 Solid Modeling: Introduction

The AutoCAD modeler in Release 13 provides solid modeling capabilities of `Regions` (2D solids) and `Solids` (3D solids). The various commands for solid modeling found in the `Solids` toolbar **(Fig 2.46)** are techniques for creating solid primitives—boxes, cylinders, spheres, and others—that can be added or subtracted from each other to form a composite object.

THE SOLIDS TOOLBAR

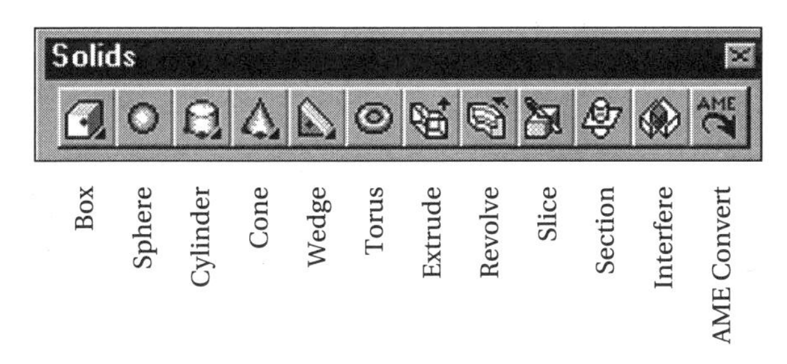

Figure 2.46 The `Solids` toolbar is used for drawing regions and solids.

DRAW TOOLBAR: 2D REGION

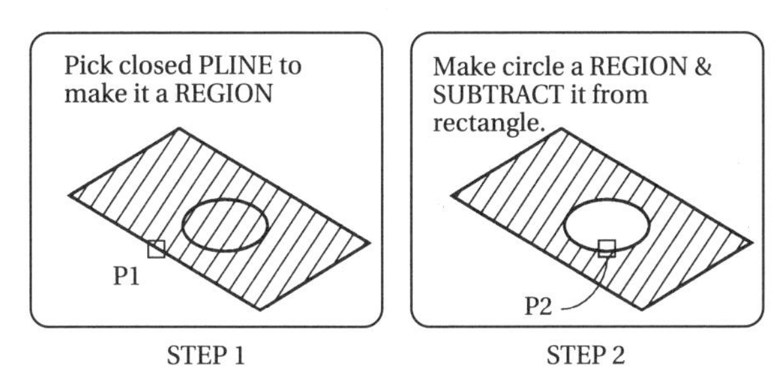

Figure 2.47 `REGION` modeling:

Step 1 `Command:` `REGION` (ENTER)
`Select objects:` (Select plane)
`1 Region created.`
Step 2 `Command:` `REGION`
`Select object:` (Select circle)
`1 Region created.`
`Command:` `SUBTRACT` (ENTER)
`Select solids and regions to subtract from...`
`Select objects:` (Select plane)
`Select objects:` (ENTER)
`Select solids and regions to subtract:`
(Select circle)
`Select objects:` (ENTER)
(Region with hole in it is created.)

REGIONS

Region modeling is a 2D version of solid modeling in which a closed surface can be converted into a solid plane (an object) **(Fig. 2.47)**. The upper plane of a wire frame enclosed by a `PLINE` is made into a 2D solid by typing `REGION` and selecting the

DRAWING TOOLBAR: 2D REGION

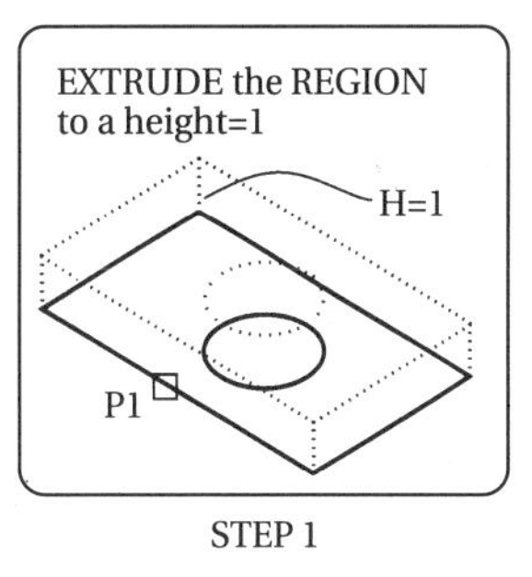

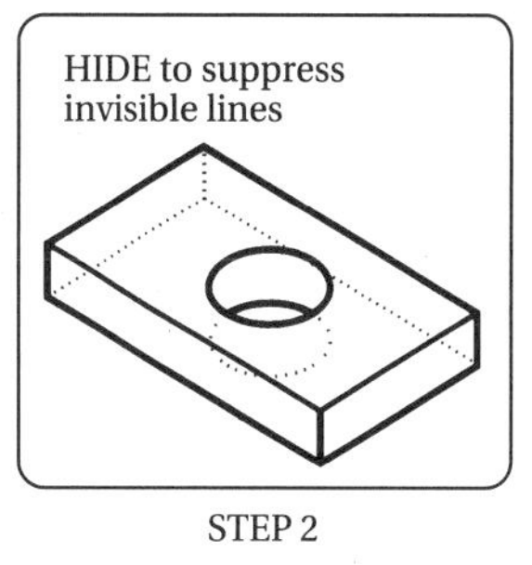

Figure 2.48 Extruding a `Region`:

Step 1 `Command:` `EXTRUDE` (ENTER)
`Select objects:` `P1` (ENTER)
Step 2 `Path/<Height of Extrusion>:` `1` (ENTER)
`Extrusion taper angle <0>:` (ENTER)
(Region is extruded.)

polyline. The circle is made into a region and is removed from the rectangular `Region` by the `SUBTRACT` command.

EXTRUDE

The `EXTRUDE` command (from the `Solids` toolbar) is used with closed polylines, polygons, circles, ellipses, and 3D entities to extrude them to a specified height, with tapered sides, if desired. **Figure 2.48** shows the extrusion of the 2D `Region` developed in Fig. 2.47 to an assigned height of 1. Polylines with crossing or intersecting segments cannot be extruded.

`Region`s can be extruded along a path (usually a polyline) to form a 3D shape as illustrated in **Fig. 2.49**. The extruded shapes can be hidden and rendered.

2.16 Extrusion Example: TILEMODE=0

Type `TILEMODE` and set it to `0` for drawing the angle bracket with the inclined surface **(Fig. 2.50)**. Begin by using `MVIEW` to create a `MS VPORT` that is about 200 × 180 to contain a single view of the bracket. Enter `Model Space (MS)` and begin drawing the bracket as follows:

DRAW TOOLBAR: EXTRUDE ON PATH

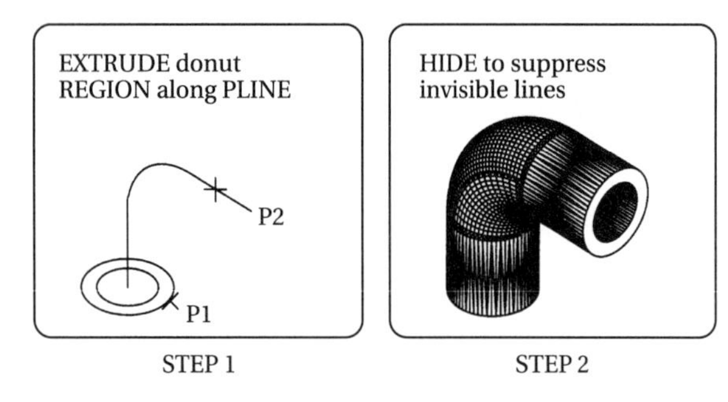

Figure 2.49 Extruding a Region:

Step 1 `Command:` EXTRUDE (ENTER)
`Select objects:` P1 (ENTER)
`Path/<Height of Extrusion>:` P (ENTER)

Step 2 `Select path:` P2 (ENTER)
(`HIDE` to show visibility.)

ANGLE BRACKET: GIVEN VIEWS

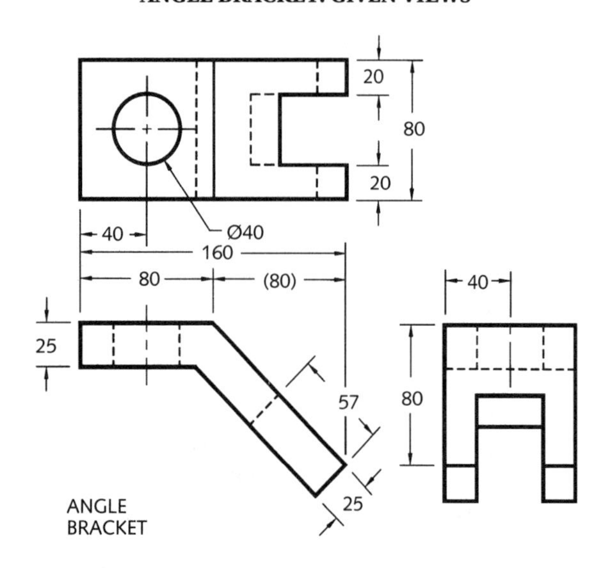

Figure 2.50 The angle bracket with an inclined surface is drawn as a 3D object in the following figures.

Part 1: Fig. 2.51 Draw the top view of the bracket as a polyline with a circle in it and set the `UCS` origin at the midpoint of one of its lines. Move the `UCSICON` to the `UCS ORigin`, obtain an isometric view (`VPOINT=1,-1,1`), and `EXTRUDE` the surface to –25 mm below the upper surface.

ANGLE BRACKET: PART 1

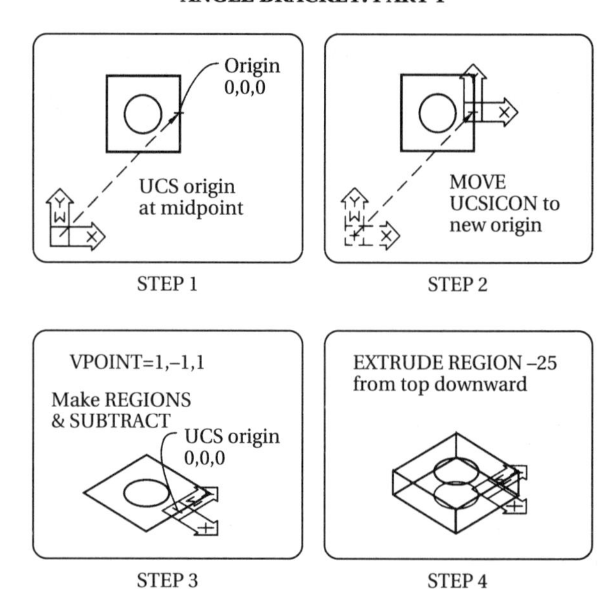

Figure 2.51 First extrusion:

Step 1 The top view of the object is drawn and the UCS origin is set at the midpoint of the line.

Step 2 `UCSICON` and `ORigin` are used to place the icon at the UCS origin.

Step 3 A `VPOINT` of `1,-1,1` is set to obtain an isometric view of the plane. Make the plane into a `Region` by subtracting the circle from it.

Step 4 `EXTRUDE` the `Region` to a height of -25.

Part 2: Fig. 2.52 Type `PS` to return to `Paper Space` and define four `VPORTS` to fill the screen to obtain four views of the bracket. Type `MS` to enter `Model Space`, activate the lower left port, and rotate the `UCS` 90° about the X-axis, making the `UCSICON` parallel to the object's front view. Type `PLAN` to obtain the front view and type `UCS` and `SAVE` to keep this `UCS` as `FRONT`.

Part 3: Fig. 2.53 `RESTORE` the `FRONT UCS` and rotate the `UCS` 90° about the Y-axis to align the `XY` icon

ANGLE BRACKET: PART 2

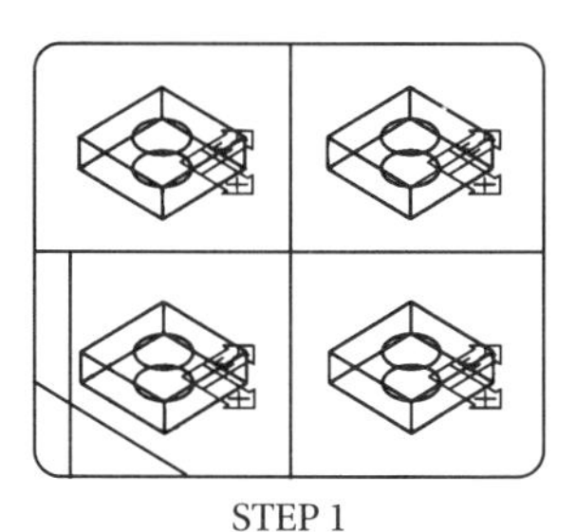

STEP 1

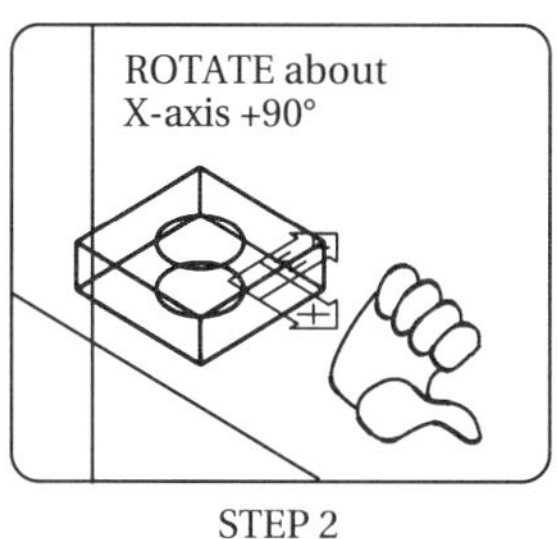

STEP 2

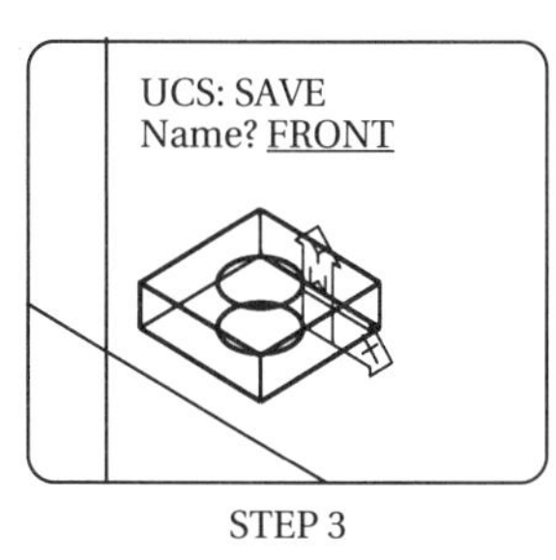

STEP 3

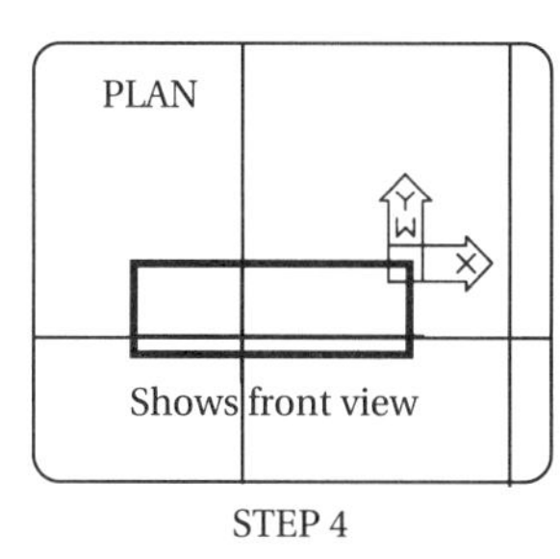

STEP 4

Figure 2.52 Four-port construction:

Step 1 `Command:` VPORTS (ENTER) (Select four ports and the 3D drawing is shown in each.)

Step 2 (Select lower left port.) `Command:` UCS (ENTER) (Select `X` option and rotate UCS 90° about the X-axis for a front view.)

Step 3 `Command:` UCS (ENTER) (Enter SAVE) `?/Desired UCS name:` FRONT (ENTER)

Step 4 `Command:` PLAN (ENTER) (Get front view and `ZOOM` to the desired size.)

ANGLE BRACKET: PART 3

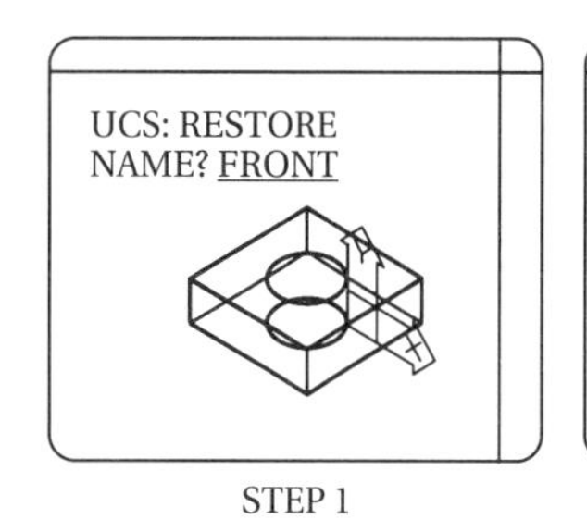

STEP 1

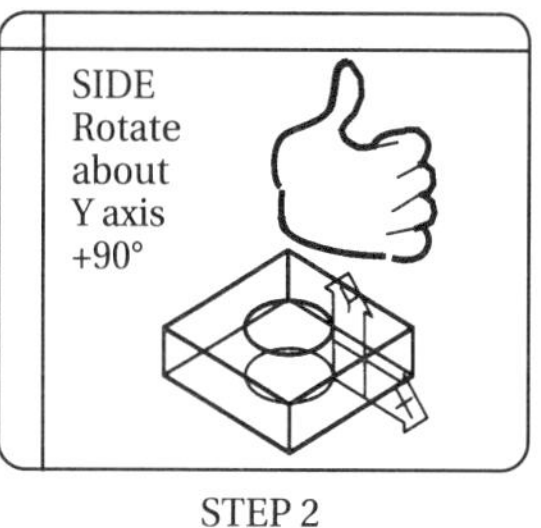

STEP 2

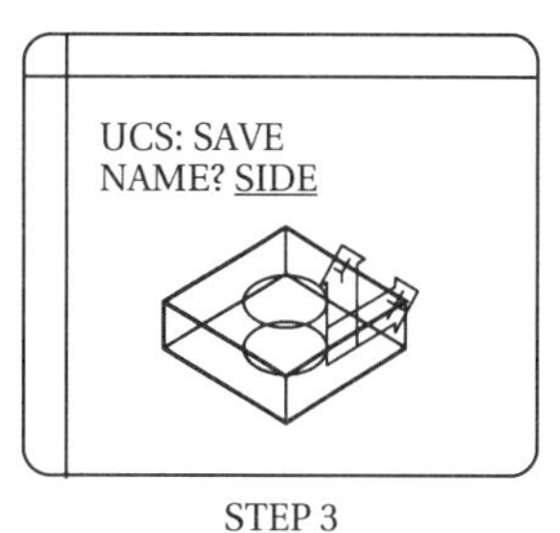

STEP 3

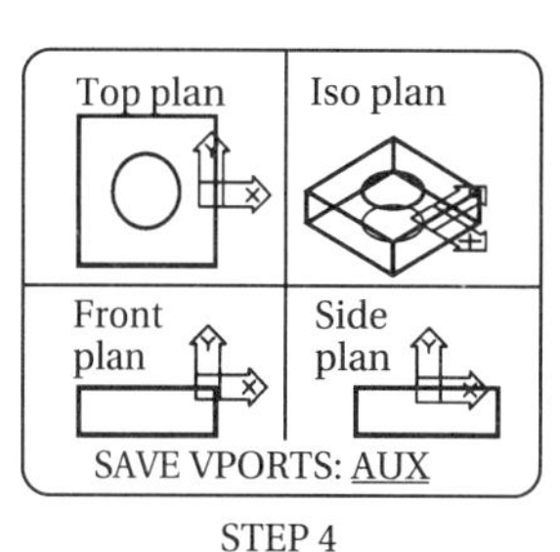

STEP 4

Figure 2.53 Top and side views:

Step 1 You may `RESTORE` the `Front` view by using the `UCS` command, but it is unnecessary here.

Step 2 (Select the lower right port.) `Command:` UCS (ENTER)

(Select Y and rotate the UCS 90°.)

Step 3 `Command:` UCS (ENTER)

`Save ?/Desired UCS name:` SIDE (ENTER)

Step 4 Command: PLAN (ENTER) (Obtain the side view.) Use the same steps to obtain and save a top view in the upper left port. Leave the 3D view in the upper right corner.

parallel to the side plane of the bracket. Activate the lower right port, `SAVE` the `UCS` as `Side`, and type `PLAN` to obtain the right-side view. Obtain and `Save` the top view in the upper right panel in the same manner as the two previous views.

Part 4: Fig. 2.54 Activate the right-side port to `Restore` the `UCS` named `Side` and show it in isometric by setting `VPOINT` to `1,-1,1`. Set the `UCS` origin with coordinates of `0,-80,-80` at the end of the sloping surface and the `UCSICON` will move to it. Rotate the `UCSICON` –45° about the X-axis to make it lie in the plane of the inclined surface. `Save` this `UCS` as `INCLINED`. Draw the notch by using the dimensions from its given views (Fig. 2.54). You may draw these points with the cursor while observing the polar coordinates at the bottom of the screen because the plane of the inclined surface lies in the `UCS`.

Part 5: Fig. 2.55 Use the `EXTRUDE` command to make a solid shape of the inclined plane with a thickness of –25. The extruded shape appears in all ports. Use `UNION` to join the two extruded shapes into a single bracket. `HIDE` can be used to remove invisible lines at this point.

ANGLE BRACKET: PART 4

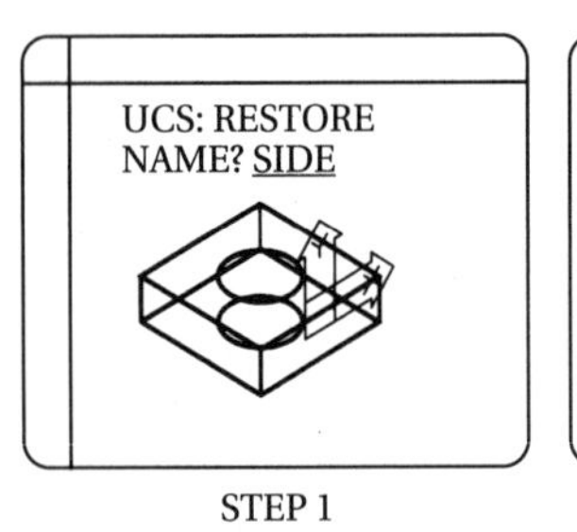

STEP 1

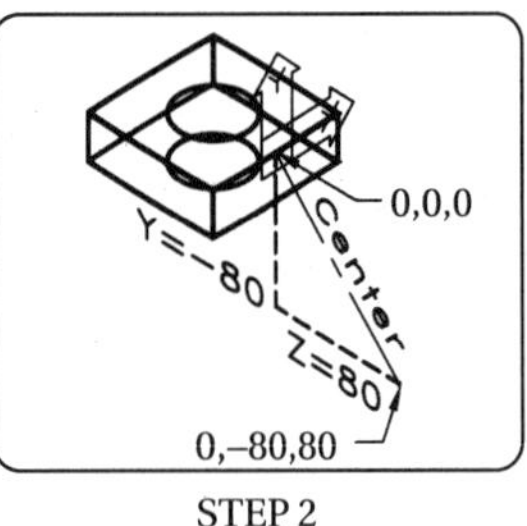

STEP 2

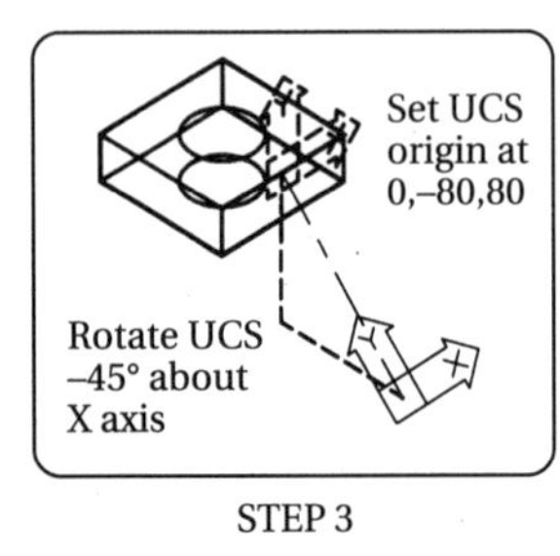

STEP 3

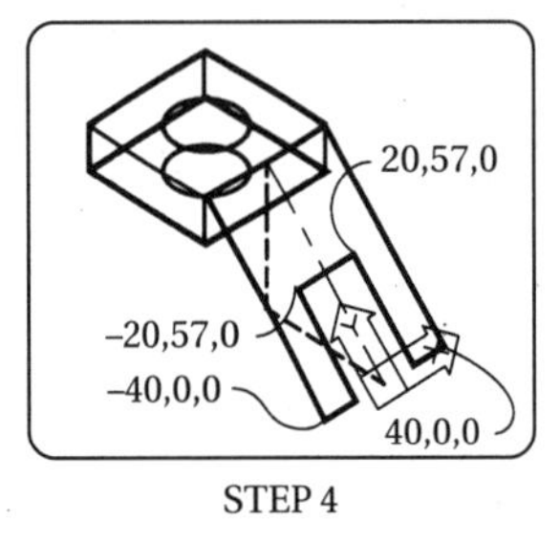

STEP 4

Figure 2.54 Drawing the inclined surface:

Step 1 (Select the side-view port.) `Command:` <u>`UCS`</u> (ENTER) (Select `RESTORE`.)
?/Name of `UCS to restore:` <u>`SIDE`</u> (ENTER)
`Command:` <u>`VIEWPORTS`</u> (ENTER) (`Save` this viewport configuration as `AUX1`.) `Command:` <u>`VPORTS`</u> (ENTER) (Select `SIngle` to obtain a full-screen image.)

Step 2 `Command:` <u>`UCS`</u> `Origin/Zaxis/... <World>:` <u>`OR`</u> (Select origin.) `Origin point:` <u>`0,-80,80`</u> (ENTER)

Step 3 `Command:` <u>`UCSICON`</u> (ENTER) (Select `ORigin` to place icon at a new origin.)
`Command: UCS` (ENTER) (Select `X` and rotate `UCS` 45° about the X-axis.) Use `UCS` command to `Save` as `INCLINED`.

Step 4 Use the `PLINE` command and draw the inclined plane; convert it into a `Region`.

2.17 Solid Primitives

Solid primitives—`box`, `sphere`, `wedge`, `cone`, `cylinder`, and `torus`—can be selected from the `Solids` toolbar (Fig. 2.46). Hidden lines in solid primitives are suppressed by typing `HIDE`.

`BOX` creates a box as shown in **Fig. 2.56** where its base is drawn in the plane of the current `UCS`. The dimensions of the box can be created with

ANGLE BRACKET: PART 5

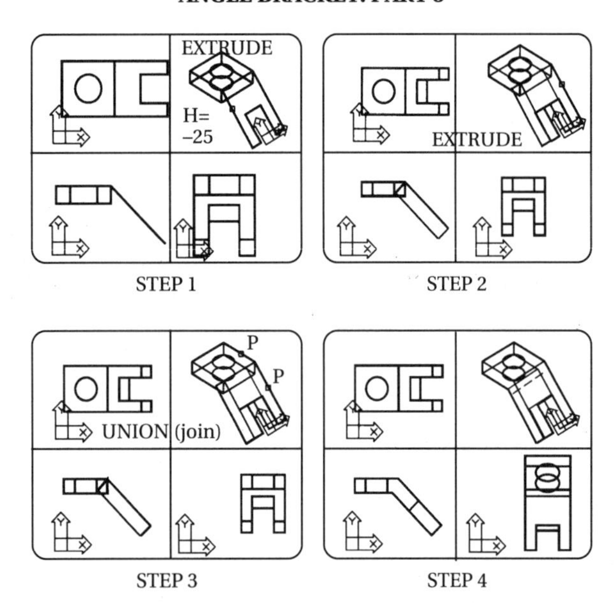

STEP 1 STEP 2 STEP 3 STEP 4

Figure 2.55 Completing the drawing:

Step 1 `Command:` <u>`VPORTS`</u> (ENTER) (`RESTORE` the four-port configuration, `AUX1`.) `EXTRUDE` the inclined plane to have a height of –25.

Step 2 The lower surface of the inclined plane is displayed.

Step 3 Use `UNION` to join the two extruded shapes.

Step 4 The object is complete and ready for the `HIDE` command.

BOX (SOLIDS TOOLBAR)

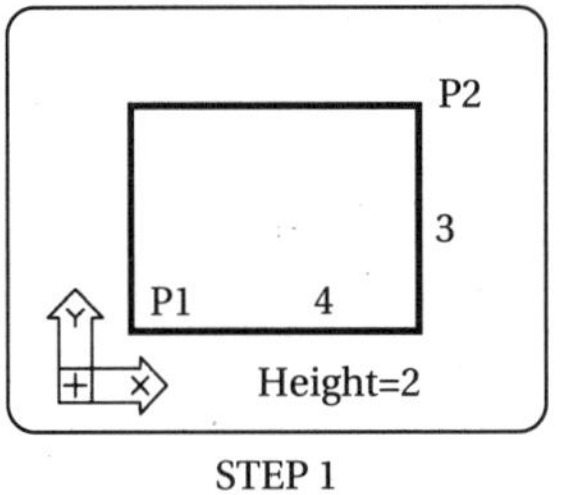

STEP 1

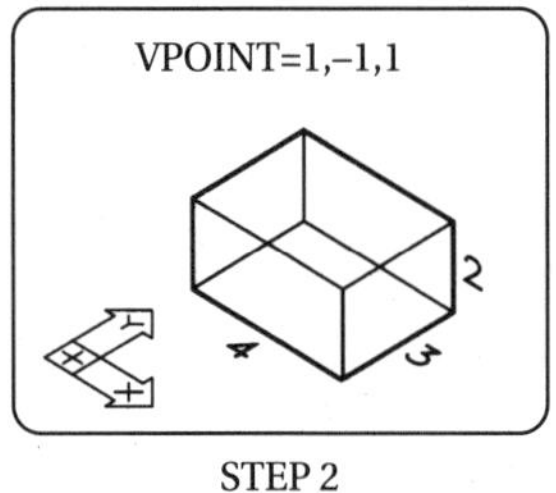

STEP 2

Figure 2.56 `BOX` command.

Step 1 `Command:` <u>`BOX`</u> (ENTER)
`Center/<Corner of box>:` <u>`P1`</u> (ENTER)
`Cube/Length/<Other corner>:` <u>`P2`</u> (ENTER)
`Height <1>:` <u>`2`</u> (ENTER)

Step 2 `Command:` <u>`VPOINT`</u> (ENTER)
`Rotate/<View point> <0,0,0>:` <u>`1,-1,1`</u> (ENTER)
(3D view of a solid box is obtained.)

SPHERE (SOLIDS TOOLBAR)

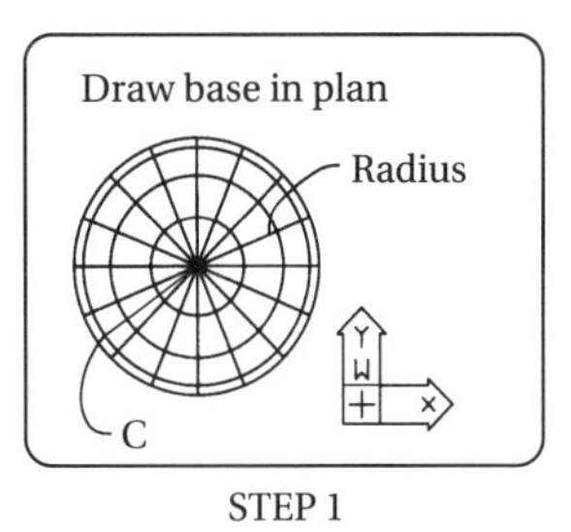

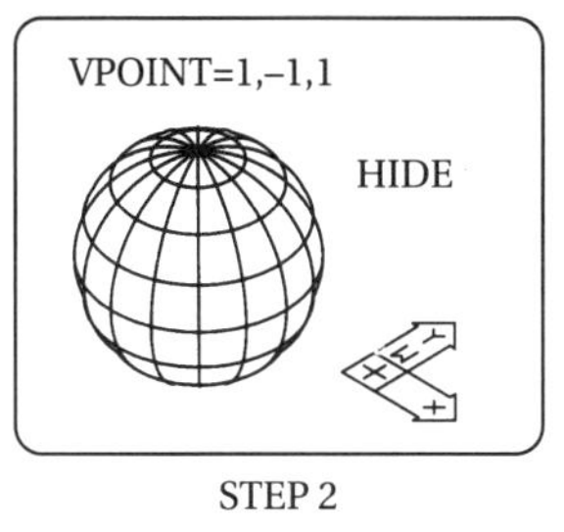

Figure 2.57 SPHERE command:

Step 1 Command: SPHERE (ENTER)
Center of sphere <0,0,0>: C (ENTER)
Diameter/<Radius> of sphere: 4 (ENTER)

Step 2 Command: VPOINT (ENTER)
Enter vpoint <0,0,0>: 1,-1,1 (ENTER)
(3D view of a sphere is obtained.)

separate widths and depths, diagonal corners of the base, or as a cube by typing values at the keyboard or selecting them with the cursor.

SPHERE creates a ball by responding to the prompts with the center and radius, or center and diameter of the ball as shown in **Fig. 2.57**. The axis of the sphere connecting its north and south poles is parallel to the Z-axis of the current UCS, and its center is on the plane of the UCS.

CYLINDER draws a cylinder in a manner similar to drawing a cone. **Figure 2.58** illustrates the steps of drawing a cylinder with a circular base. Cylinders with elliptical bases also may be drawn.

CONE draws cones with circular or elliptical bases, as shown in **Fig. 2.59** with a circular base by locating its center, axis endpoints, and height.

WEDGE draws the base of a wedge that lies in the plane of the current UCS with the upper plane sloping toward the second point selected **(Fig. 2.60)**. Prompts ask for the length, width, and height of the wedge.

CYLINDER (SOLIDS TOOLBAR)

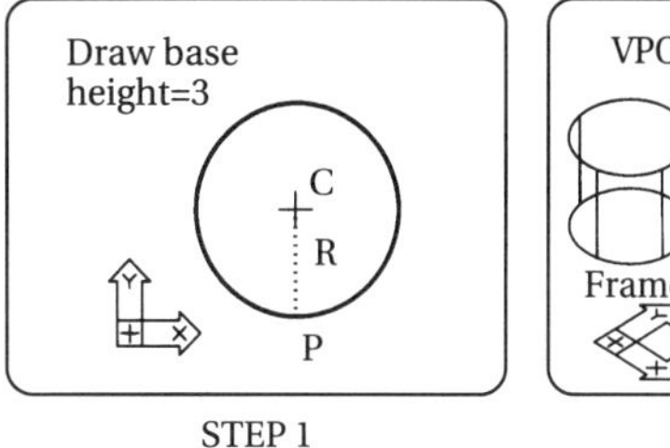

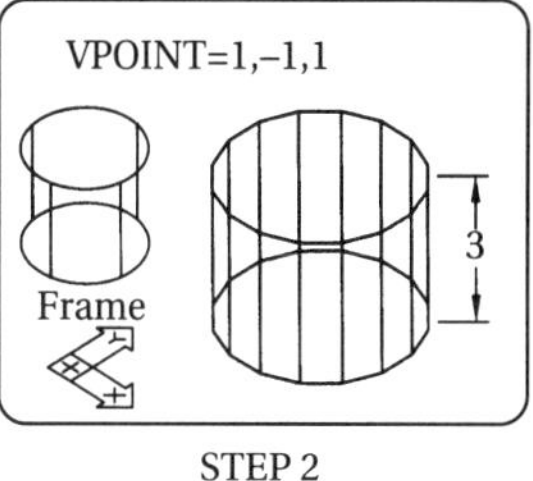

Figure 2.58 CYLINDER command:

Step 1 Command: CYLINDER (ENTER)
Elliptical/<Center point>:C (ENTER)
Center of other end/<Height>: 3 (ENTER)

Step 2 Command: VPOINT (ENTER)
Rotate/<View point> <0,0,0>: 1,-1,1 (ENTER)
(3D view of a cylinder is obtained.)

CONE: CIRCULAR (SOLIDS TOOLBAR)

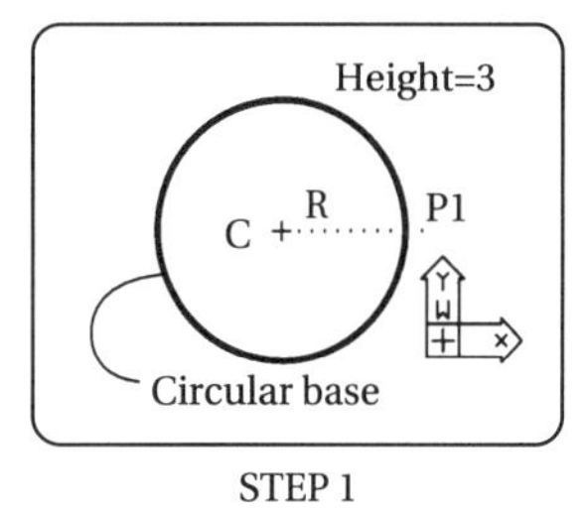

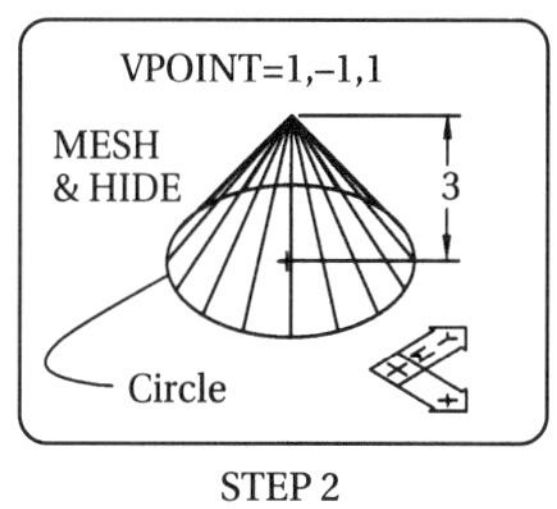

Figure 2.59 CONE command (circular base):

Step 1 Command: CONE (ENTER)
Elliptical/<Center point> <0,0,0>: C (ENTER)
Apex/<Height>: 3 (ENTER)

Step 2 Command: VPOINT (ENTER)
Rotate/<View point> <0,0,0>: 1,-1,1 (ENTER)

TORUS draws a donut solid by giving its center, diameter or radius of the tube, and diameter or radius of revolution **(Fig. 2.61)**. The diameter of the torus will lie in the plane of the current UCS.

REVOLVE is used to sweep polylines, polygons, circles, ellipses, and 3D poly objects about an

WEDGE (SOLIDS TOOLBAR)

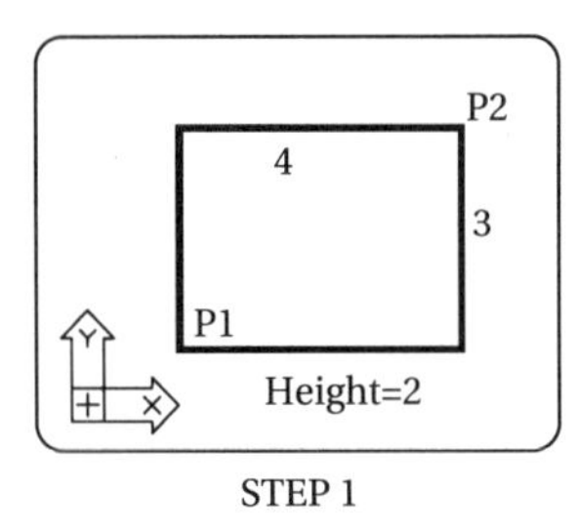

STEP 1

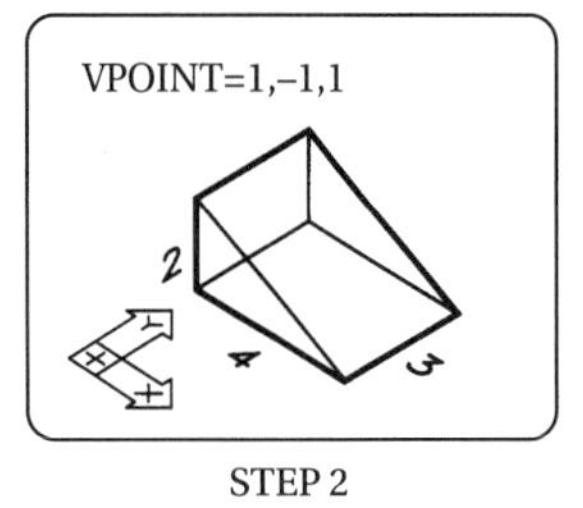

STEP 2

Figure 2.60 WEDGE command:

Step 1 Command: WEDGE (ENTER)

Center/<Corner of wedge> <0,0,0>: P1 (ENTER)

Cube/Length/<Other corner>: P2 (ENTER)

Height: 2 (ENTER)

Step 2 Command: VPOINT (ENTER)

Rotate/<View point> <0,0,0>: 1,-1,1 (ENTER)

TORUS (SOLIDS TOOLBAR)

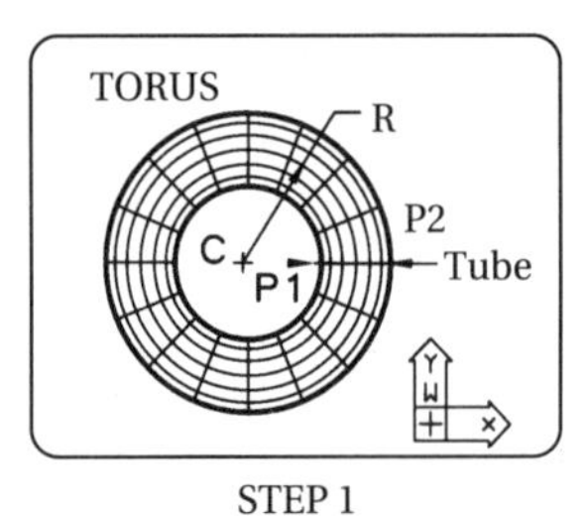

STEP 1

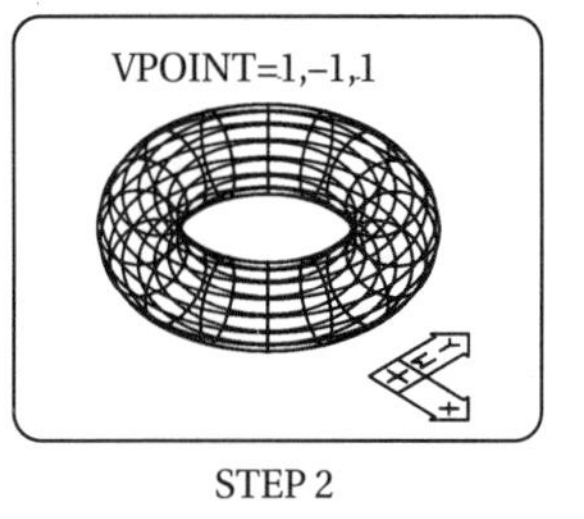

STEP 2

Figure 2.61 TORUS command:

Step 1 Command: TORUS (ENTER)

Center of torus: C (ENTER)

Diameter/<Radius> of torus: (Select.) (ENTER)

Diameter/<Radius> of tube: (Select.) (ENTER)

Step 2 Command: VPOINT (ENTER)

Rotate/<View point> <0,0,0>: 1,-1,1 (ENTER)

(3D view of a solid torus is obtained.)

axis if they have at least 3, but less than 300, vertices. In **Fig. 2.62**, a polyline is revolved a full 360° about an axis. The path of revolution can start and end at any point between 0 and 360°. Polylines that have been Fit or Splined will require extensive calculations.

REVOLVE (SOLIDS TOOLBAR)

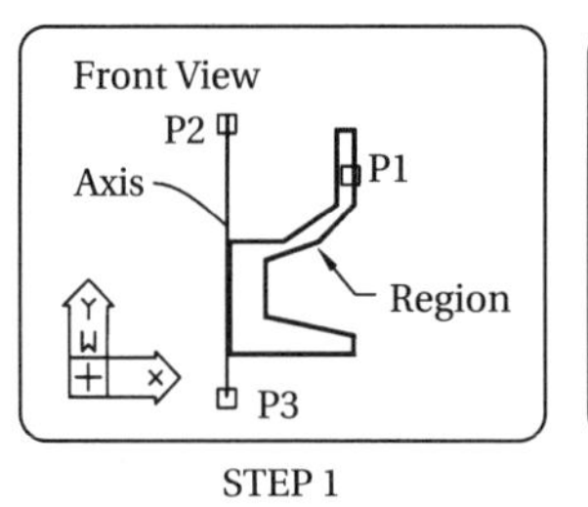

STEP 1

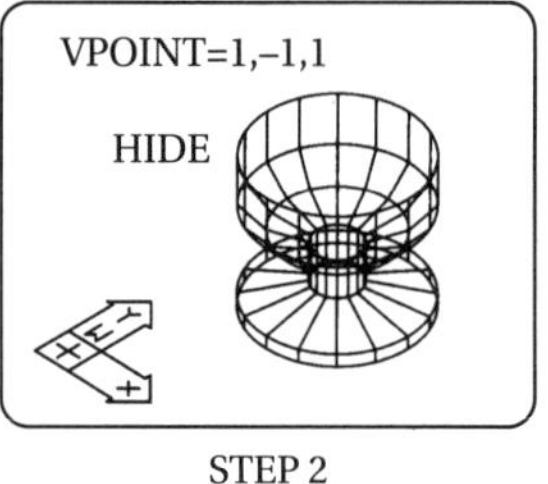

STEP 2

Figure 2.62 REVOLVE command:

Step 1 Command: REVOLVE (ENTER)

Select objects: P1 (ENTER) (Select region.)

Axis of revolution -Object/X/Y/<Start point of axis>: P2

<End point of axis>: P2

Angle of revolution <full circle>: (ENTER)

(Surface is revolved.)

Step 2 Command: VPOINT (ENTER)

Rotate/<View point> <0,0,0>: 1,-1,1 (ENTER)

(3D view of the revolved solid is obtained.)

2.18 Modifying Solids

Once Solids have been drawn several modification commands can be used to refine them: SUBTRACT, UNION, EXPLODE, CHAMFER, FILLET, EXTEND, and TRIM.

SUBTRACT is used to remove one intersecting solid from another as illustrated in **Fig. 2.63**.

UNION is used to join intersecting solids to form a single composite solid model. **Figure 2.64** shows how a box and cylinder are unified into a single solid.

INTERFERE is used to obtain a solid that is common to two intersecting solids. An interference solid is found in **Fig. 2.65** with this command.

EXPLODE is used to separate solids or regions that were combined by the SUBTRACT and UNION commands to permit editing or correcting before redoing SUBTRACT and UNION commands.

SUBTRACT (MODIFY TOOLBAR)

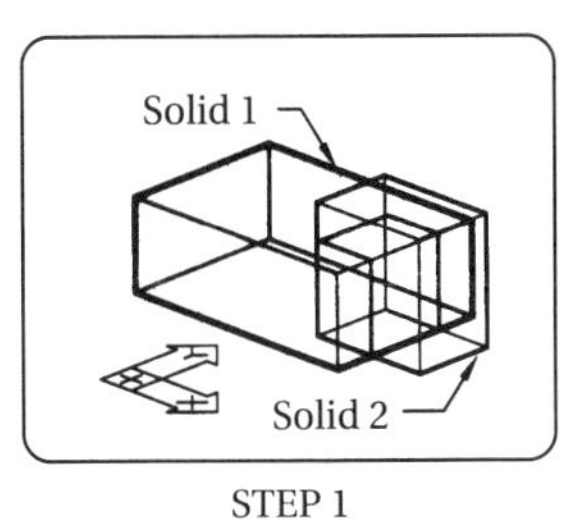

STEP 1

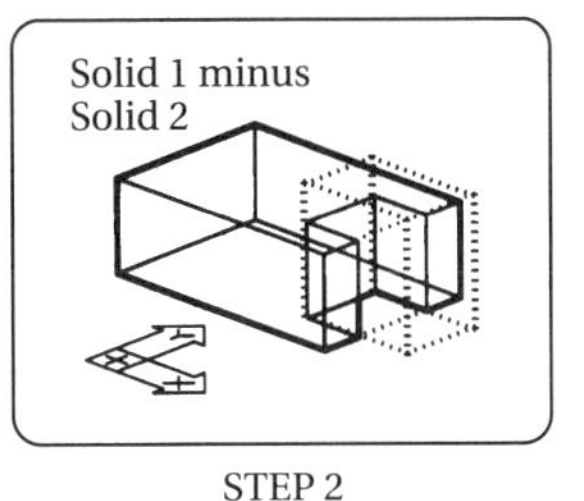

STEP 2

Figure 2.63 SUBTRACT command:

Step 1 Command: <u>SUBTRACT</u> (ENTER)
Select solids and regions to subtract from... <u>P1</u> (ENTER)
Select solids and regions to subtract... <u>P2</u> (ENTER)
Select objects: (ENTER)

Step 2 Command: <u>VPOINT</u> (ENTER)
Rotate/<View point> <0,0,0>: <u>1,-1,1</u> (ENTER)
(3D view of the resulting solid is obtained.)

UNION (MODIFY TOOLBAR)

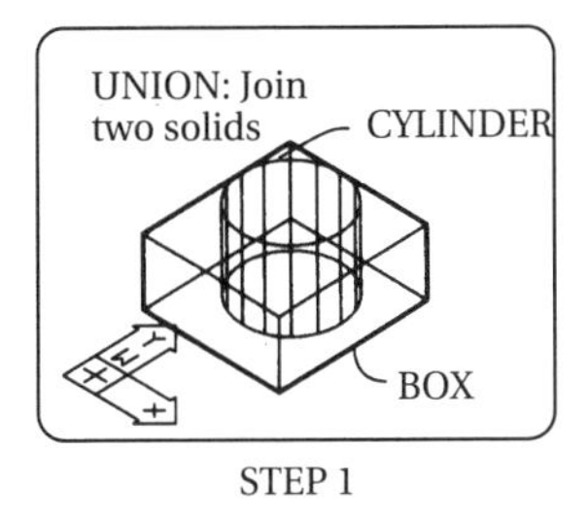

STEP 1

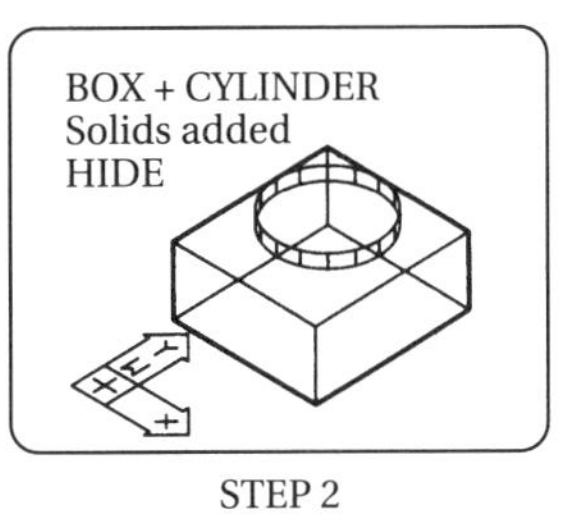

STEP 2

Figure 2.64 UNION command:

Step 1 Command: <u>UNION</u> (ENTER)
Select objects: (Select box.) (ENTER)
Select objects: (Select cylinder.) (ENTER)
Select objects: (ENTER)

Step 2 Command: <u>VPOINT</u> (ENTER)
Rotate/<View point> <0,0,0>: <u>1,-1,1</u> (ENTER)
(3D view of the composite solid is obtained.)

CHAMFER applies beveled edges by selecting the base surface, the adjoining surface, and the edges to be chamfered, giving the first and second chamfer distances **(Fig. 2.66)**. When

INTERFERE (MODIFY TOOLBAR)

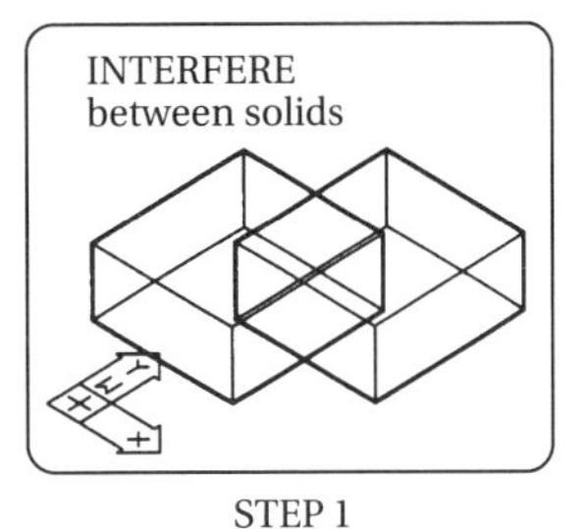

STEP 1

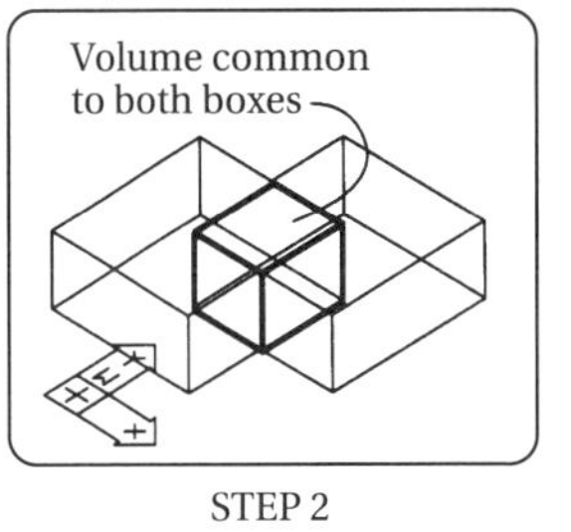

STEP 2

Figure 2.65 INTERFERE command:

Step 1 Command: <u>INTERFERE</u> (ENTER)
Select the first set of solids:
Select objects: <u>P1</u> (ENTER)
Select second set of solids:

Step 2 Select objects: <u>P2</u> (ENTER)
Create interference solids ?<N>: <u>Y</u> (ENTER)
(Interference solid is created.)

CHAMFER (MODIFY TOOLBAR)

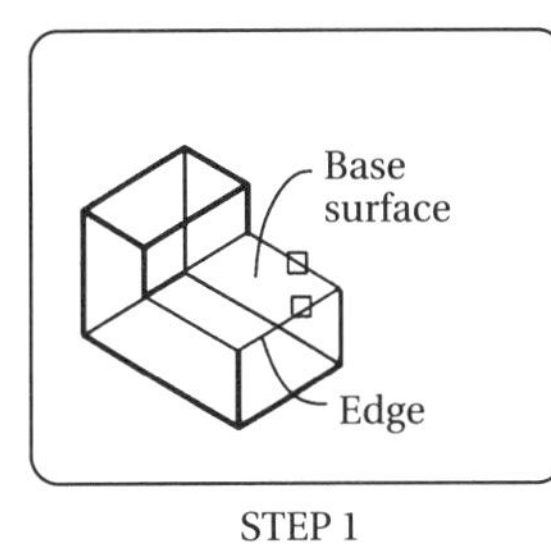

STEP 1

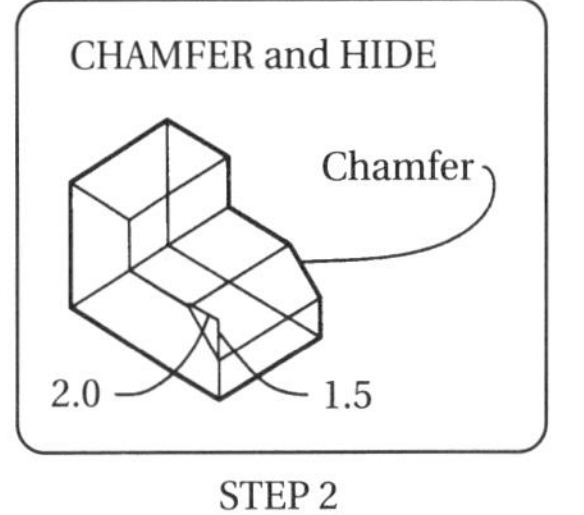

STEP 2

Figure 2.66 CHAMFER command:

Step 1 Command: <u>CHAMFER</u> (ENTER)
Select base surface: (Select.)
<OK>/ Next: (ENTER)
Select edges to be chamfered (Press (ENTER) when done): (Select.) (ENTER)
1 edge selected
Enter distance along first surface <default>: <u>2</u>
Enter distance along second surface <default>: <u>1.5</u>

Step 2 Command: <u>VPOINT</u> (ENTER)
Enter vpoint <0,0,0>: <u>1,-1,1</u> (ENTER)
(3D view of chamfered object is obtained.)

FILLET (MODIFY TOOLBAR)

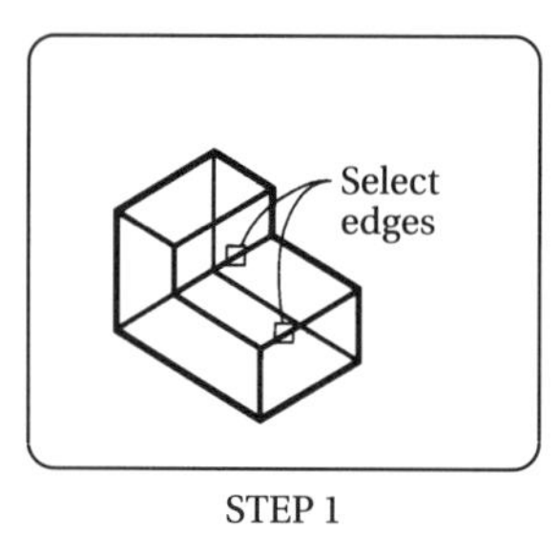

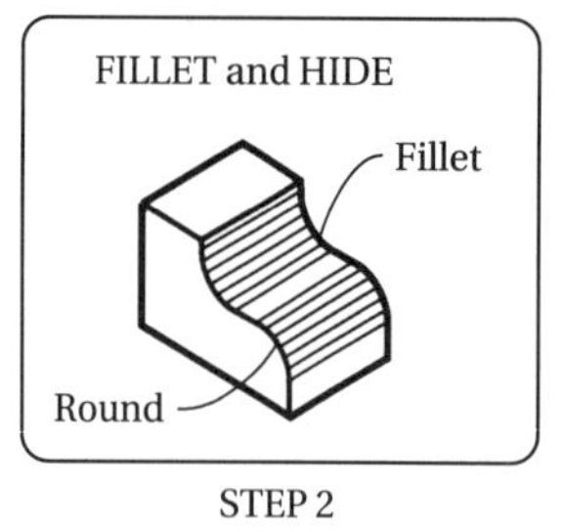

Figure 2.67 `FILLET` command:

Step 1 `Command:` <u>`FILLET`</u> (ENTER)
`Select edges to be filleted (Press` (ENTER) `when done):` (Select 3 edges.) (ENTER)
`3 edges selected`
`Diameter/<Radius> of fillet <default>:` <u>`3`</u> (ENTER)

Step 2 `Command:` <u>`VPOINT`</u> (ENTER)
`Enter vpoint <0,0,0>:` <u>`1,-1,1`</u> (ENTER)
(3D view of filleted object is obtained.)

SECTION (SOLIDS TOOLBAR)

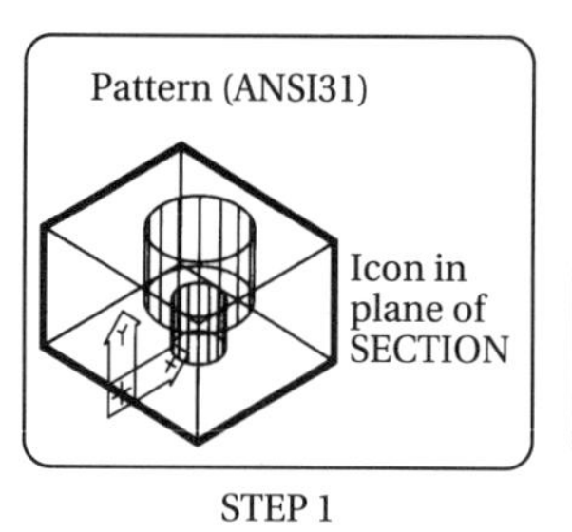

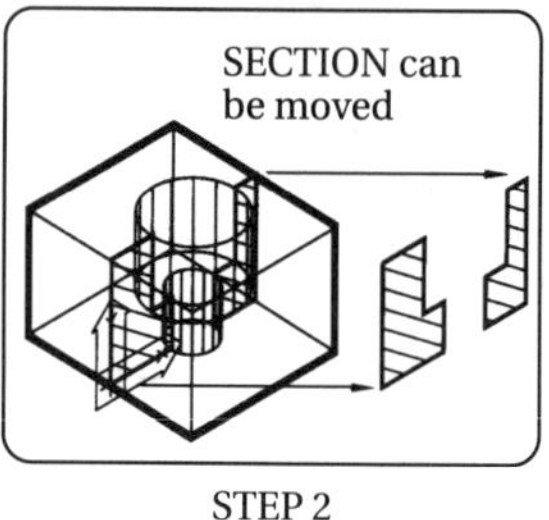

Figure 2.68 `SECTION` command:

Step 1 (Position the UCS in the plane of the section.)
`Command:` <u>`SECTION`</u> (ENTER)
`Select objects:` (Select box.) (ENTER)
`Section plane by Object/Zaxis/View/ XY/YZ/ZX/<3points>:` <u>`XY`</u> (ENTER)
(The section establishes a `Region` through the part.)

Step 2 `Command:` <u>`MOVE`</u> (ENTER)
(Move the `Region` outside the part; apply section lines to it if you like.)

these prompts have been satisfied, the chamfers are automatically drawn.

`FILLET` is used to apply rounded intersections between planes by selecting the edges of solids and giving the diameter or radius of the fillet, as shown in **Fig. 2.67**. The `FILLET` command can also be used to fillet the edge of a cylindrical feature.

2.19 SECTION

`SECTION` is used to pass a sectioning plane through a 3D solid to show a `Region` that outlines its internal features. `BHATCH` (hatch pattern) can be used to assign the hatching pattern to the `Region` if it lies in the plane of the `UCS`. In **Fig. 2.68**, the hatch pattern is set to `ANSI31` for a cast iron symbol.

The `UCSICON` is placed on the object to establish the plane of the section, `SECTION` is typed, and the plane passing through the object appears. The section plane can be moved as shown in this example. Other sections through the object are found in this same manner by positioning the icon in the cutting plane or by selecting from one of the following options: `3point`, `Object`, `Zaxis`, `View`, `XY`, `YZ`, or `ZX`.

2.20 SLICE

With the `SLICE` command an object can be cut through and made into separate parts, either or both of which can be retained as shown in **Fig. 2.69**. The `SLICE` command has the following options:

`3points` defines three points on the slice plane.

`Object` aligns the cutting plane with a circle, ellipse, 2D spline, or 2D polyline element.

`Zaxis` defines a plane by picking an origin

SLICE (SOLIDS TOOLBAR)

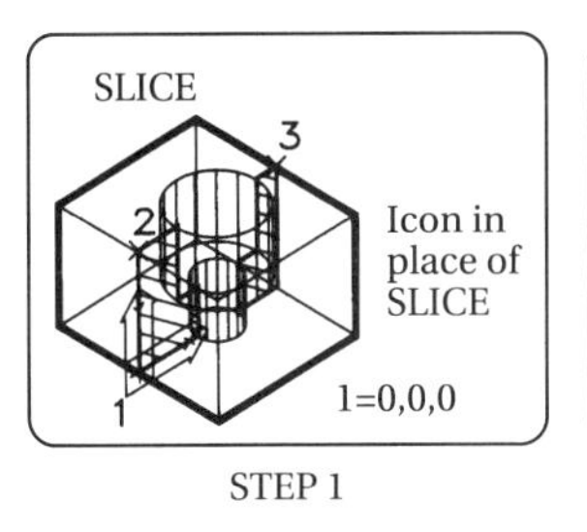

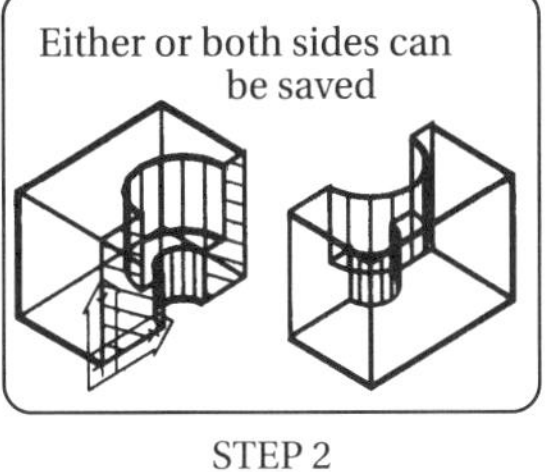

Figure 2.69 SLICE command:

Step 1 (Move the UCS to the plane of the SLICE.)

Command: SLICE (ENTER)

Select objects: (Pick the object)

Slicing plane by Object/Zaxis/View/XY/ YZ/ZX/<3Points>: XY (ENTER)

Step 2 Point on XY plane <0,0,0>: 2 (ENTER)

Both sides/<Point on desired side of plane>: B (ENTER)

BRACKET

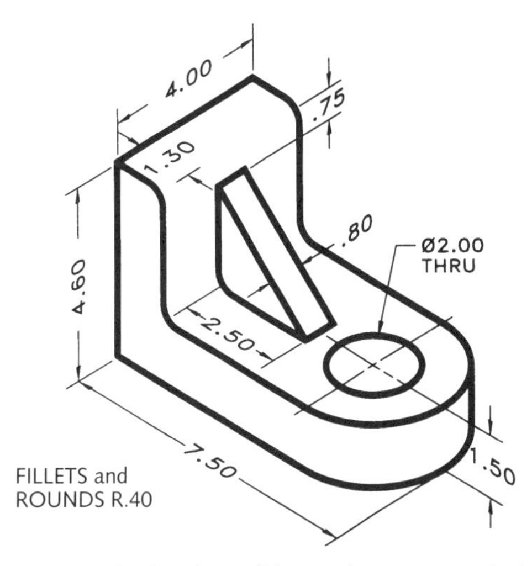

Figure 2.70 This bracket will be used as an example for illustrating solid modeling in the following figures.

point on the Z axis that is perpendicular to the selected points.

View makes the cutting plane parallel with the viewport's viewing plane when a single point is selected.

XY, YZ, or ZX aligns the cutting plane with the planes of the UCS by selecting only one point.

2.21 A Solid Model Example

The bracket shown in **Fig. 2.70** is to be drawn as a solid model by using the commands from the Solids toolbar. Crate the plan view of the base as a PLINE and obtain a isometric view of it by VPOINT 1,-1,1 **(Fig. 2.71)**. Then, EXTRUDE the base to a height of 1.50 units.

A BOX of 1.30 x 4.00 x 3.10 is drawn **(Fig. 2.72)** and moved to the base by SNAPping End P1 of the box End A. Use WEDGE to draw the bracket's rib as shown in **Fig. 2.73**. Move the rib to join the base and the upright box by using the Midpoint option of SNAP.

BRACKET: EXTRUDE

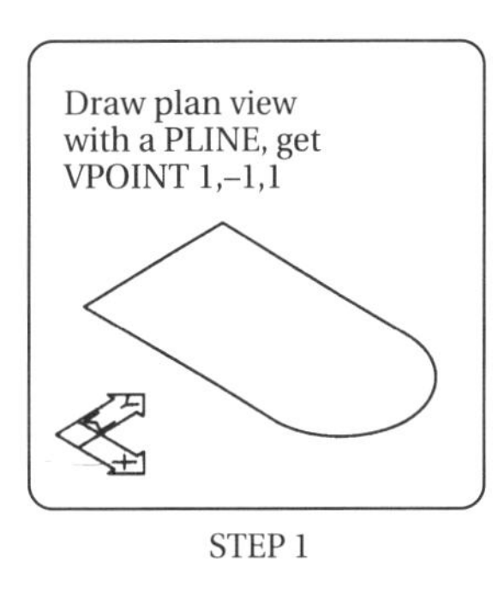

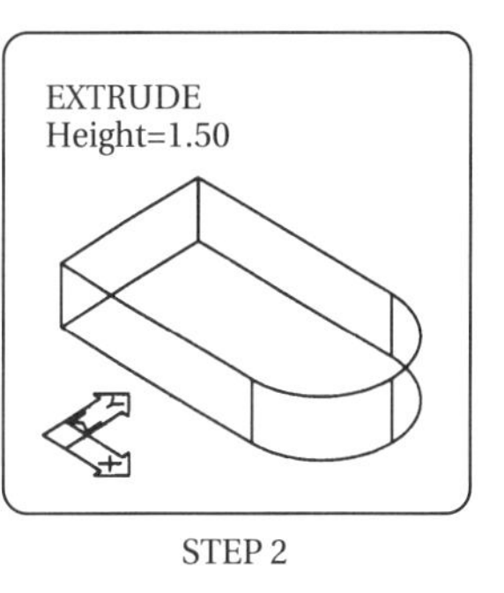

Figure 2.71 Bracket (EXTRUDE command):

Step 1 Draw plan view of base of bracket with a PLINE command.

Command: VPOINT (ENTER)

Rotate/<Viewpoint>: 1,-1,1 (ENTER)

(Get isometric view of base.)

Step 2 Command: EXTRUDE (ENTER)

Select objects: (Select PLINE.) (ENTER)

Path/<Height of Extrusion>: 1.50 (ENTER)

Extrusion taper angle <0>: (ENTER)

(Base is extruded 1.50 high.)

BRACKET: BOX

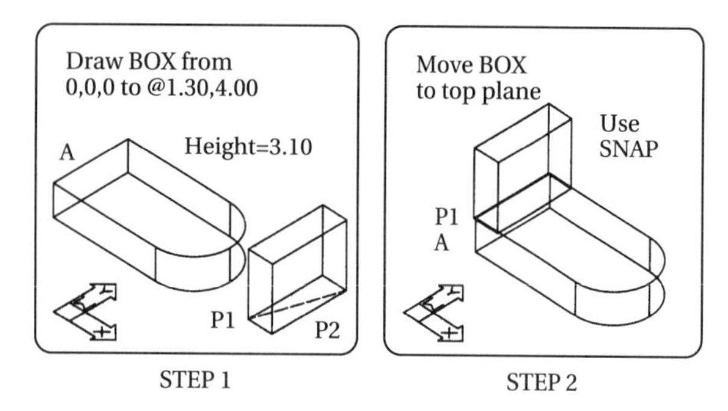

Figure 2.72 Bracket (BOX command):

Step 1 Command: BOX ENTER
Center/<Corner of box><0,0,0>: ENTER
Cube/Length/<Other corner>: @1.30.4 ENTER
Height: 3.1 ENTER (Box is drawn.)

Step 2 (Set SNAP to END.)
Command: MOVE ENTER
Select objects: (Select box.)
Base point or displacement: P1 ENTER
Second point of displacement: A ENTER
(Box is moved to base extrusion.)

BRACKET: WEDGE

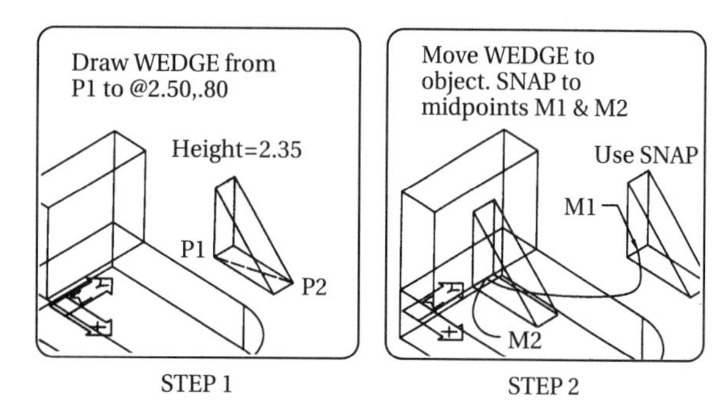

Figure 2.73 Bracket (WEDGE command):

Step 1 Command: WEDGE ENTER
Center/<Corner of wedge>: P1 ENTER
Length/<Other corner>: @2,5,.8 ENTER
Height: 2.35 ENTER (Wedge is drawn.)

Step 2 (Set snap to mid.)
Command: MOVE ENTER
Select objects: (Select wedge.)
Base point or displacement: M1 ENTER
Second point of displacement: M2 ENTER
(Wedge is moved to midpoint of line.)

BRACKET: CYLINDER

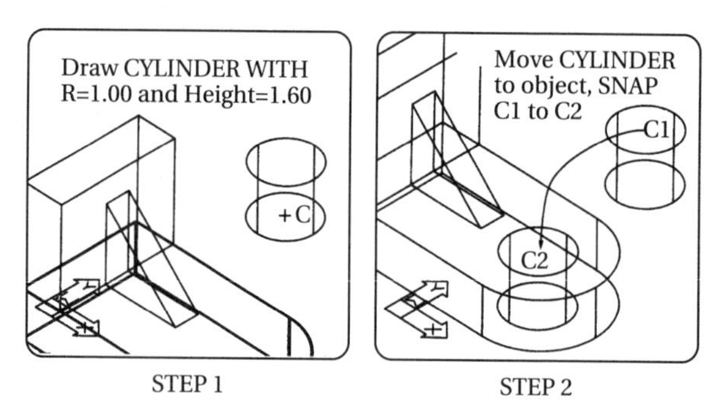

Figure 2.74 Bracket (CYLINDER command):

Step 1 Command: CYL (or CYL) ENTER
Elliptical/<Center point> <0,0,0>: C
Diameter/<Radius>: 1.00 ENTER
Center of other end/<Height>: 1.60 ENTER
(Cylinder is drawn.)

Step 2 (Set SNAP to CENT.)
Command: MOVE ENTER
Select objects: (Select cylinder.)
Base point or displacement: P1 ENTER
Second point of displacement: P2 ENTER
(Cylinder is moved to center of arc.)

In **Fig. 2.74**, use CYLINDER to represent the hole and MOVE it to the center of the semicircular end of the base using the Center option of SNAP. Use SUBTRACT to create the hole in the base **(Fig. 2.75)**. Use the UNION command to join the base, upright box, and wedge together into a composite solid.

FILLET is used in **Fig. 2.76** to select the edges to be rounded with a radius of 0.40 and HIDE suppresses the invisible lines **(Fig. 2.77)**. An infinite number of views of the bracket can be obtained with VPOINT or DVIEW.

2.22 MASSPROP (Mass Properties)

Various properties of regions and solids can be obtained with the MASSPROP command:

Area computes the area of a region or solid.

Perimeter calculates the perimeter of a region; not available for solids.

BRACKET: SUBTRACT AND UNION

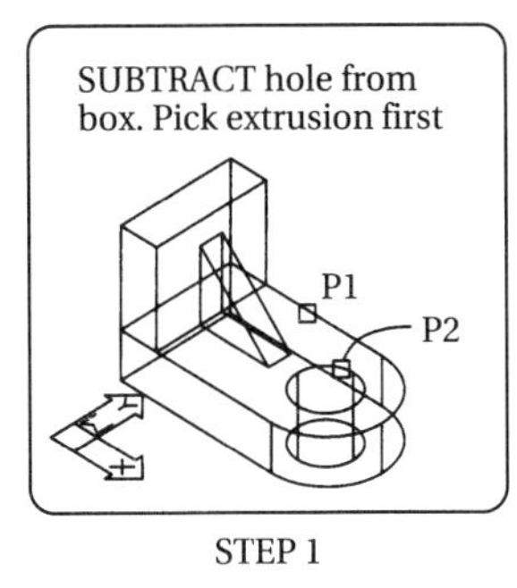

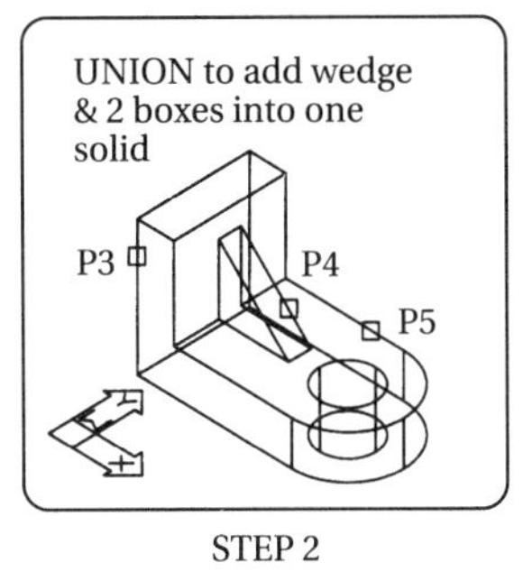

Figure 2.75 Bracket (`SUBTRACT` and `UNION` commands):

Step 1 `Command:` <u>`SUBTRACT`</u> (ENTER)
`Select solids and regions to subtract from...`
`Select objects:` <u>`P1`</u> (ENTER)
`Select solids and regions to subtract...`
`Select objects:` <u>`P2`</u> (ENTER)
(Hole is subtracted.)

Step 2 `Command:` <u>`UNION`</u> (ENTER)
`Select objects:` <u>`P3, P4, P5`</u> (ENTER)
`Select objects:` (ENTER)
(Bracket is unified.)

BRACKET: FILLET

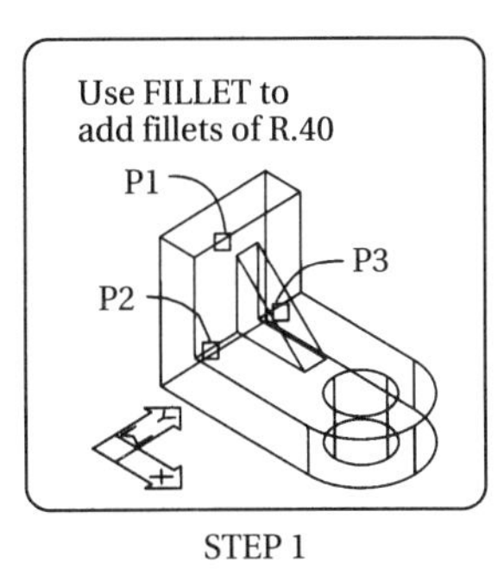

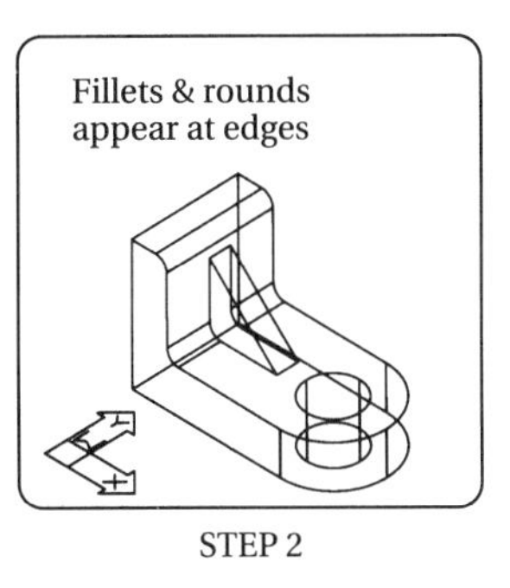

Figure 2.76 Bracket (`FILLET` command):

Step 1 `Command:` <u>`FILLET`</u>
`(Trim mode) current fillet radius=0.00`
`Polyline/Radius/Trim/<Select first object>:` <u>`R`</u> (ENTER)
`Enter fillet radius <0.00>:` <u>`.40`</u> (ENTER) (ENTER)

Step 2 `Command:` <u>`FILLET`</u> (ENTER)
`Polyline/Ra... <Select first object>:` <u>`P1`</u> (ENTER)
`Enter radius <.40>:` (ENTER)
`Chain/Radius/<Select edge>:` <u>`P1, P2, P3`</u> (ENTER)

BRACKET: HIDE

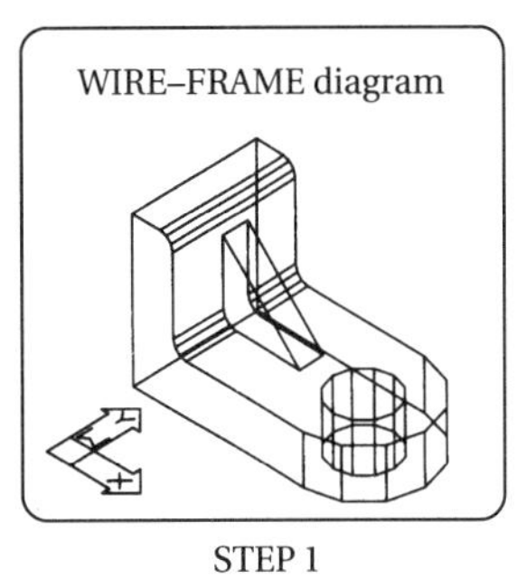

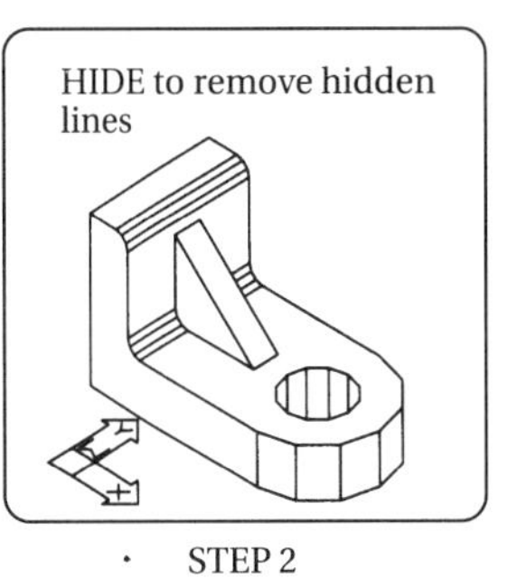

Figure 2.77 Bracket (`HIDE` command):

Step 1 `Command:` <u>`HIDE`</u> (ENTER)

Step 2 (The object is shown as a solid with hidden lines removed.)

`Bounding Box` gives coordinates of the diagonal corners of a region's enclosing rectangle. For a solid, coordinates of the diagonal and opposite corners of a 3D box are given.

`Volume` gives the 3D space enclosed in a solid.

`Mass` gives the weight of a solid.

`Centroid` gives the coordinates of the center of a region or the 3D center of a solid.

`Moment of Inertia` is given for regions and solids.

`Product of Inertia` is given for regions and solids.

`Radius of Gyration` is given for regions and solids.

`MATLIB` (materials library) assigns material types that can be assigned to solids that will have an effect on the mass properties listed above. The `Materials Library` box is found on the `Render` toolbar.

2.23 Paper Space and Model Space: TILEMODE=0

By setting `TILEMODE=0 (Off)`, we can plot all VPORTS just as they appear on the screen, not just

TILEMODE=0: PAPER SPACE

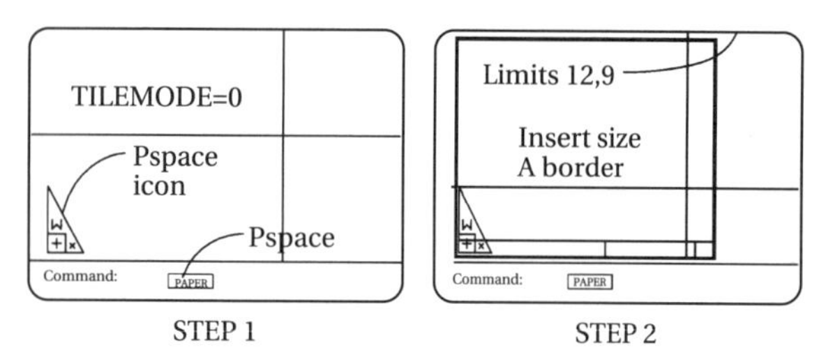

Figure 2.78 `Tilemode=0`: `PSPACE` option:

Step 1 `Command:` `TILEMODE` ENTER
`New value for TILEMODE <1>: 0` ENTER (Enters `PSPACE`; 2D icon and `P` in status line appear.)

Step 2 `Command:` `INSERT` ENTER
`Block name (or?):` `BORDER-A` ENTER
`Insertion point:` `0,0` ENTER
`X scale factor <1>/Corner/XYZ:` `1` ENTER
`Y scale factor (default = X):` `2` ENTER
`Rotation angle <0>:` `0` ENTER (Border is inserted in `Pspace`. Set limits to `12, 9` and `ZOOM/ALL`.)

MVIEW TO CREATE MS WINDOWS

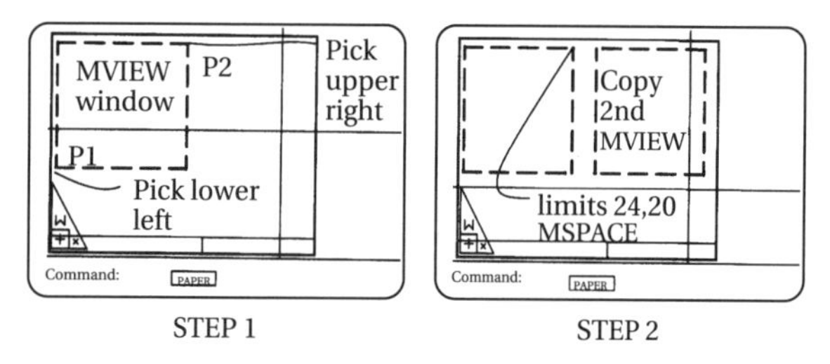

Figure 2.79 `Tilemode=0`: `MVIEW` option:

Step 1 `Command:` `MVIEW` ENTER (Switching to paper space.)
`ON/OFF/Hideplot/Fit/2/3/4/Restore/<First Point>:` `P1`
`Other corner:` `P2` ENTER (A 3D model-space port is drawn.)

Step 2 While still in paper space, make a copy of the 3D model-space port, return to `MS`, and set `VPORT` limits to `24, 20`.

the active `VPORT`. (An overview of `Paper Space` and `Model Space` was covered in Section 2.2.)

Type `UCSICON` and `On` so the icon will appear. Type `TILEMODE` and set to `0` (`Off`) to turn the screen

DRAWING IN MODEL SPACE (MS)

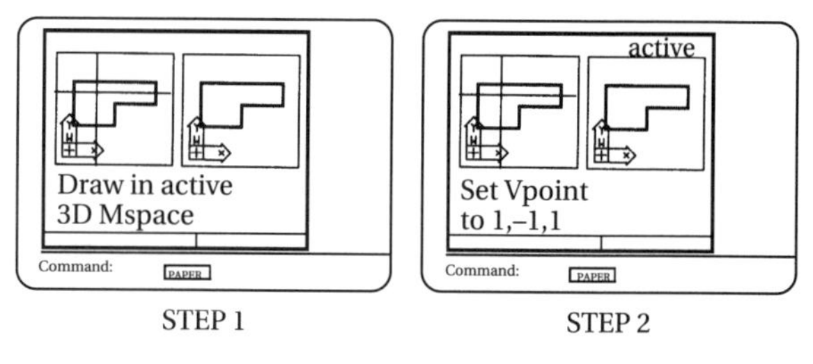

Figure 2.80 `Tilemode=0` (Drawing in model space):

Step 1 `Command:` `MSPACE` (3D icons appear in both ports and the cursor is in the active port.)

(Draw an extruded part and it will show in both views.)

Step 2 Pick the right viewport to make it active.
`Command:` `VPOINT` ENTER
`Rotate/<View point> <current>:` `1,-1,1` ENTER (An isometric view is obtained.)

into `Paper Space` (blank screen) with a `PSPACE` icon (a triangle) in its lower left corner, and `PAPER` appears in the status bar **(Fig. 2.78)**. Then set the paper space `Limits` large enough to contain the border, about 12" × 9", and insert an A-size border.

Type `MVIEW`, select diagonal corners of a model-space viewport with the cursor, `COPY` it, type `MSPACE` (`MS`) to enter `Model Space`, and make one of the model spaces active **(Fig. 2.79)**. Set the `Limits` in this 3D viewport to `24, 20`, large enough to contain the drawing. Move between `MS` and `PS` by double clicking on the button in the status bar that will be either `MODEL` or `PAPER`.

Other options under `MVIEW`, which operates from `Paper Space`, are `On`, `Off`, `Hideplot`, `Fit`, `2`, `3`, `4`, and `Restore`.

`On/Off`: Selects and turns off viewports to save regeneration time, but leave at least one on.

`Hideplot`: When turned on, removes the hidden lines from the selected viewports.

`Fit`: Makes a viewport fill the current screen.

`2/3/4`: Lets you create an area and specify the number of viewports within it.

SCALING WITH ZOOM

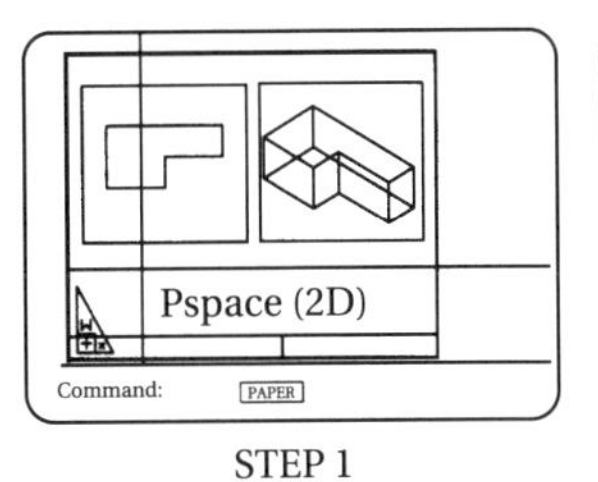

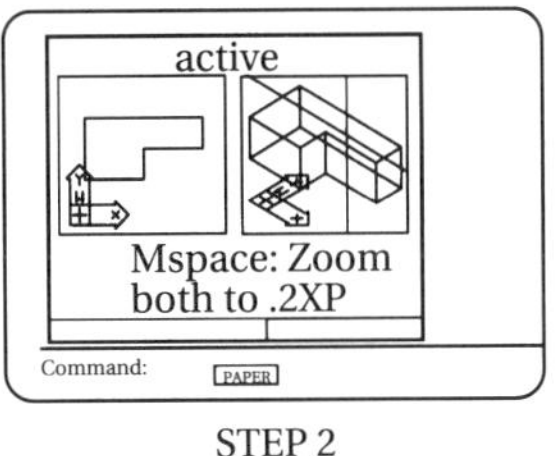

Figure 2.81 `Tilemode=0` (`ZOOM`ing to scale):

Step 1 `Command:` `PSPACE` (ENTER) (The 2D border reappears and the cursor spans the screen. Since `PS` had limits of `12, 9`, and `MS` had limits of `24, 20`, the `MS` ports must be `ZOOM`ed to about 0.20 size for both to fit in border.)

Step 2 `Command:` `MSPACE` (ENTER) (One of the 3D ports becomes active.)

`Command:` `ZOOM` (ENTER)

`All/Center/... /<Scale (X/XP)>:` `.2XP`

(Active port is scaled to 0.2 size to fit within 2D border. Select and `ZOOM` other view to `0.2XP`.)

MOVING MS VIEWPORTS

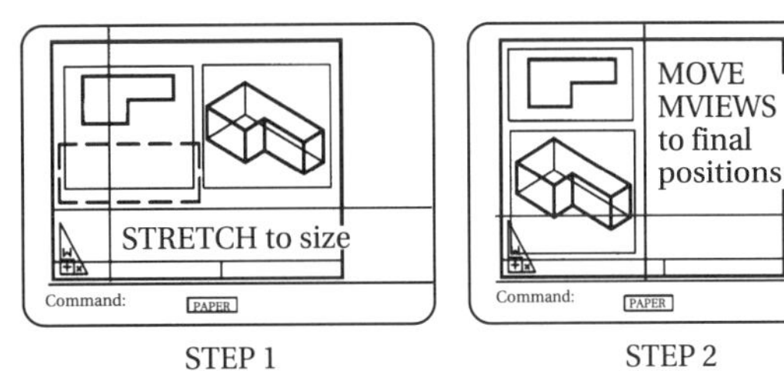

Figure 2.82 `Tilemode=0` (Moving ports):

Step 1 `Command:` `PSPACE` (ENTER) (Switch to `PS`. Use `STRETCH` to size the 3D ports.)

Step 2 `MOVE MS` ports to their final positions within the 2D border in paper space.

PLOTTING FROM PAPER SPACE (PS)

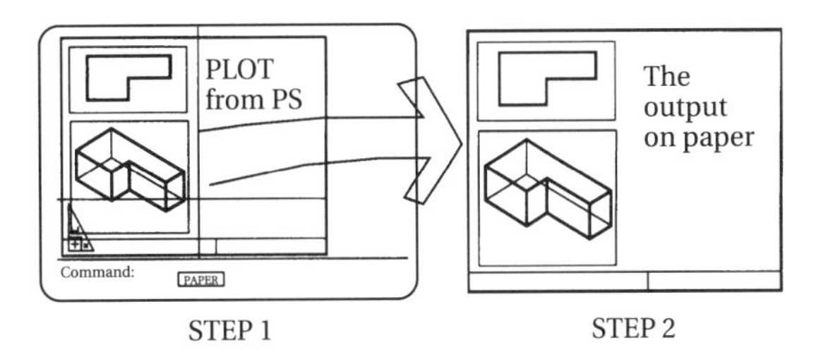

Figure 2.83 `Tilemode=0` (Plotting):

Step 1 `Command:` `PLOT` (ENTER) (Give specification for plotting from paper space.)

Step 2 The A-size layout is plotted to show both 2D (paper space) and 3D (model space) in the same plot.

`Restore:` Recalls a viewport configuration `Save`d by `VPORTS`.

Figure 2.80 shows a 3D object drawn by the `EXTRUDE` command in one of the 3D viewports and shown simultaneously in a second port. Select an isometric `VPOINT` and type `PS` to enter paper space. The cursor spans the screen in paper space. Now the drawing must be scaled.

The paper-space `Limits` are `12, 9`, inside of which are two model-space viewports each with `Limits` of `24, 20`. These `MS` ports must be sized to fit, which requires calculations. The combined width of the two 24-in.-wide model spaces is 48. When scaled to half size, their width is 24, too wide to fit. If scaled to 0.20 (two-tenths), their width is 9.6, small enough to fit inside the A-size border.

From model space, type `ZOOM` and use the `X/XP` (`2/10XP` or `0.2XP`) option to size the contents of each viewport **Fig. 2.81**). This factor changes the width limit of each viewport from 24 to 4.8, with both drawings having the same scale on the screen. In other words, you may scale viewports with width limits of 24 inches to a full-size width of 4.8 in paper space.

Type `PS` to enter paper space and use `STRETCH` to reduce the size of the `MSPACE` outlines **(Fig. 2.82)**. Reposition the model-space views for plotting with `MOVE`.

Plotting must be performed from `PSPACE`. Both the 2D and 3D drawings are plotted at the same time, including the outlines of the model-space ports **(Fig. 2.83)**. To remove the window outlines, `CHANGE` them to a separate `Layer` (`WIN-`

DIMENSIONING IN MS: HORIZONTAL PLANE

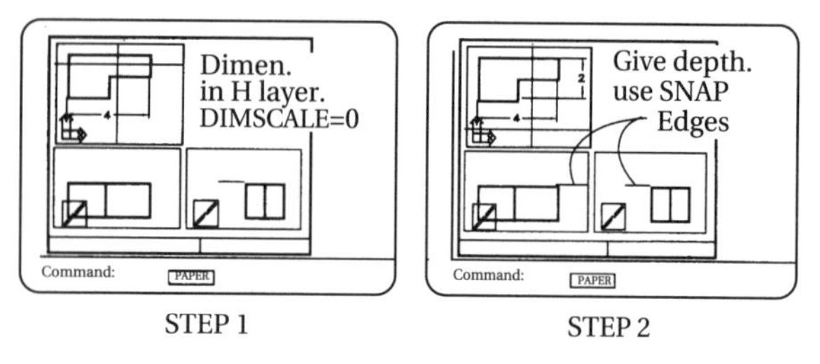

Figure 2.84 Dimensioning in model space:

Step 1 Create layer `H` and set it on. Set the `Dim Vars`, `DIMSCALE`, to 0, and the dimension variables in model space will match those specified for paper space. Move the `UCSICON` to the plane of the dimension and apply the dimensions by snapping to the endpoints of the object.

Step 2 When a dimension is applied in one viewport, it is shown in all 3D viewports and the dimensioning variables are the same size in `MS` as in `PS`. The dimensions are edges in the front and side views.

`DOW`, for example), `FREEZE` it, and `PLOT` in the usual manner.

Remove hidden lines when plotting by selecting the `Hideplot` option of `MVIEW` from `PS`, and pick the `Remove hidden lines` button from the `PLOT` dialog box. When prompted by `Hideplot`, pick the outlines of the model-space viewports. Use `Vplayer` to select 3D viewports from paper space in which layers can be turned `Off` or `Frozen` while remaining `On` or `Thawed` in other viewports.

2.24 Dimensioning in 3D

Dimensions can be applied to objects in model space to match the dimensioning variables set for paper space by setting `DIMSCALE=0` **(Fig. 2.84)**. Create two new layers, H and F, on which horizontal and frontal dimensions will be placed. Set associative dimensions on (`DIMASO=On`), set the layer H on, set the `UCS` to line in the top plane (see `XY` icon), and `SNAP` to the endpoints of the object to attach width and depth dimension to the top view. These dimensions appear as edges in the front and side views.

DIMENSIONING IN MS: FRONTAL PLANE

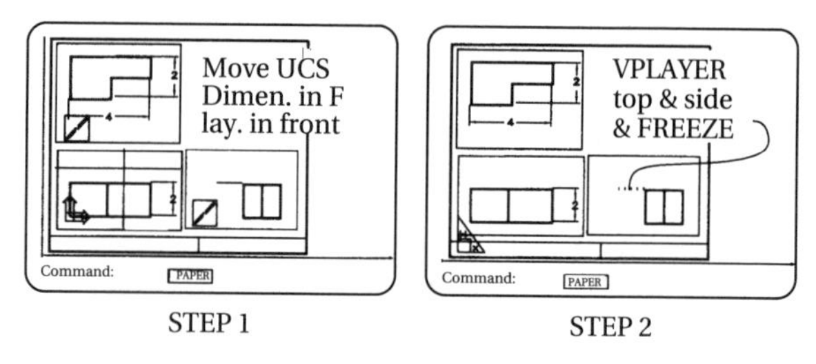

Figure 2.85 Dimensioning a 3D view:

Step 1 Create layer `F` and set it on. Move the `UCSICON` to the plane of the frontal dimension in the front view. Apply the dimensions.

Step 2 The dimension appears as an edge in the top and front views.

Select the front viewport and rotate the `XY` icon parallel to the frontal plane (`UCS/X/90<dg>`) **(Fig. 2.85)**. Attach a vertical height dimension to the object by `SNAP`ping to the endpoints of the front view and the dimension appears as an edge in the top and side view.

Since the vertical dimensions cannot be seen in the top and side views, select these ports, type `VPLAYER`, and `Freeze` the F layer **(Fig. 2.86)**. Since the horizontal dimensions cannot be seen in the front and side views, select these ports, type `VPLAYER`, and `Freeze` the H layer. Other dimensions can be added and edited in this manner.

Options under `Vplayer` that operate from paper space are `Freeze`, `Thaw`, `Reset`, `Newfrz`, and `Vpvisdflt`.

> `Freeze/Thaw`: Allows specified layers in selected `VPORTS` to be frozen or thawed. `Thaw` does not work on globally frozen layers.
>
> `Reset`: Changes visibility of one or more layers in selected `Vports` to their current default setting.
>
> `Newfrz`: Creates a new layer that is visible in the current viewport and is frozen in the rest.
>
> `Vpvisdflt`: Used to set default visibility by viewport for any layer. The default setting

VPLAYER COMMAND

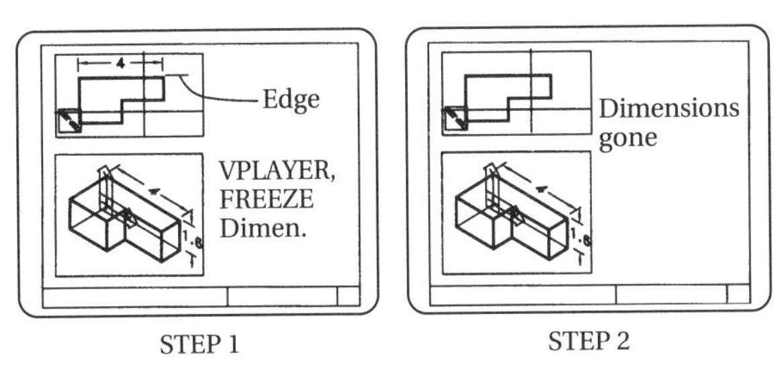

Figure 2.86 Dimensioning with the `Vplayer` option:
Step 1 Since the frontal dimension is not readable in the top view, set `VPlayer` to `On`, select the border of the 3D viewport (while in `PS`), and `Freeze` the F layer in that view.
Step 2 The dimensions are removed in the top view but remain in the 3D view. `Freeze` the H layer in the front and side view in this manner also.

BRACKET TO BE RENDERED

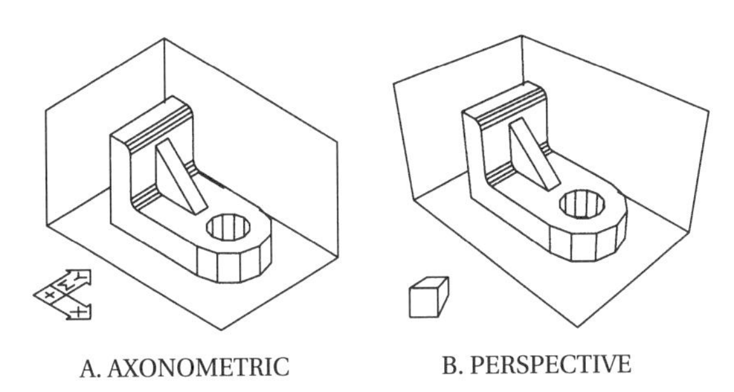

Figure 2.87 This part is used to illustrate rendering techniques in the following examples.

makes layers frozen or thawed in new viewports.

2.25 Render

Rendering is the process of adding color, lighting, and materials to 3D objects. The following examples illustrate how 3D drawings are rendered to give them a realistic appearance.

The bracket from Fig. 2.70 is displayed in two viewports as an axonometric and a perspective,

RENDER TOOLBAR

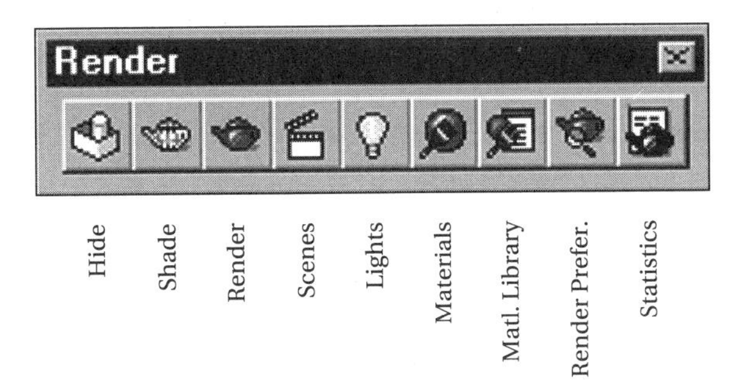

Figure 2.88 The `Render` toolbar provides the necessary commands for rendering 3D models.

RENDERING PREFERENCES DIALOG BOX

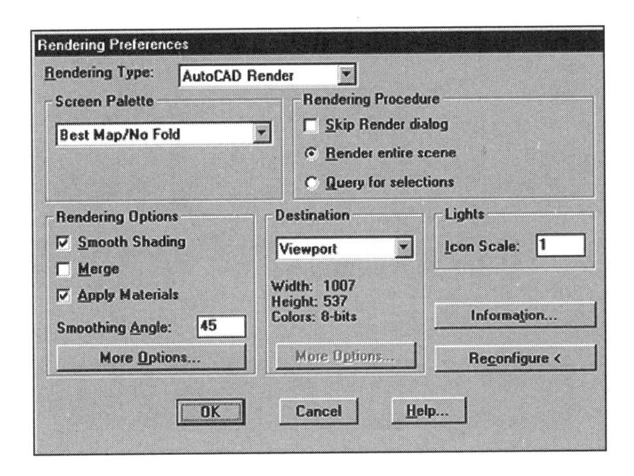

Figure 2.89 This `Rendering Preferences` dialog box is accessed from the `Render` toolbar.

respectively **(Fig. 2.87)**. From the `Render` toolbar **(Fig. 2.88)**, select `Render Preferences` to obtain the dialog box shown in **Fig. 2.89**. Rendering options are `Smooth Shading`, `Merge`, and `Apply Materials`. The `More Options` button **(Fig. 2.90)** gives `Gouraud` and `Phong` options to specify rendering quality. The `Phong` option gives the smoother, more realistic rendering.

While a view of the bracket is on the screen, select `Render` from the `Render` toolbar, or type

AUTOCAD RENDER OPTIONS (RENDER PREFERENCES)

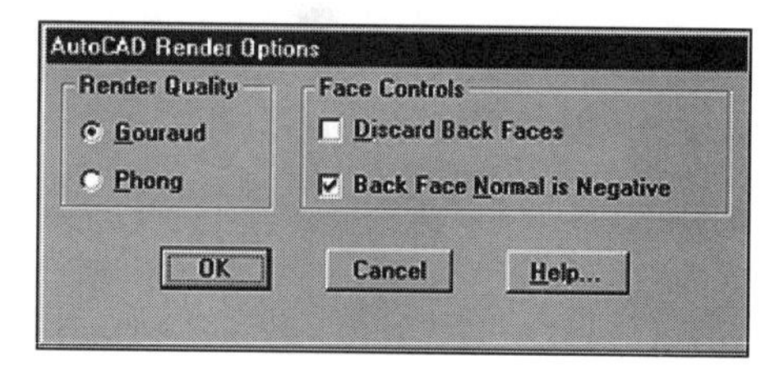

Figure 2.90 The `More Options` button is activated from the `Rendering Preferences` dialog box.

RENDER DIALOG BOX

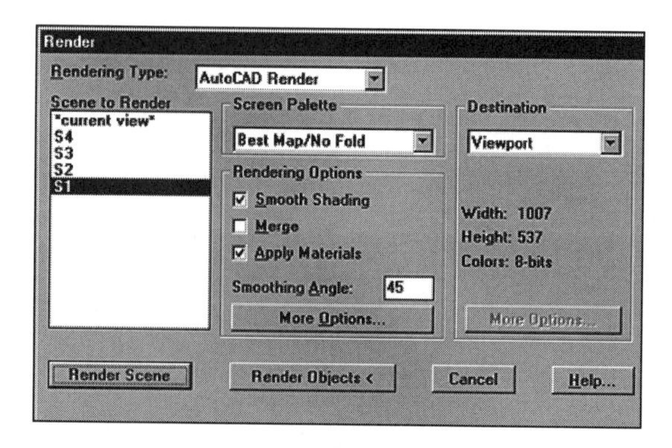

Figure 2.91 This is the `Render` dialog box selected from the `Render` toolbar.

RENDERED BRACKET

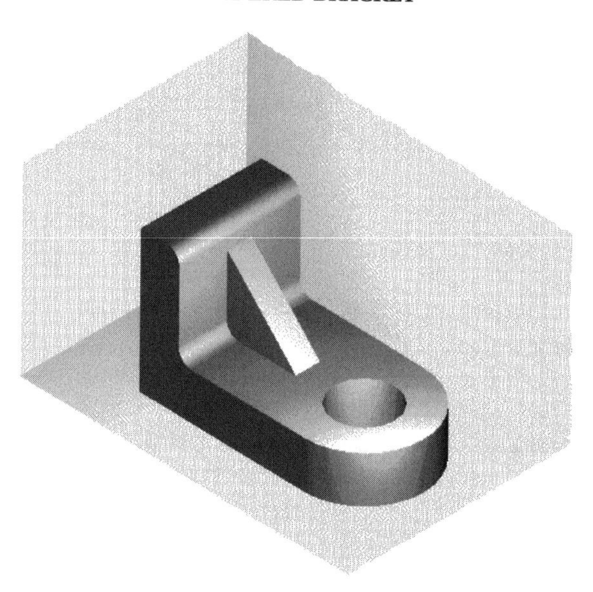

Figure 2.92 The sample rendered bracket.

LIGHT SYMBOLS

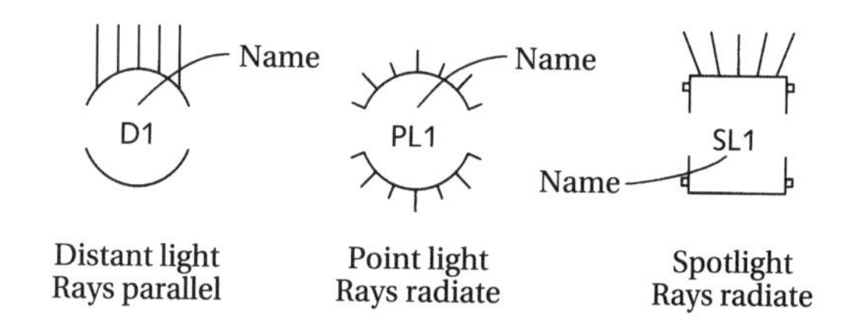

Figure 2.93 Distant light, point light, and spotlight symbols are placed on drawings to indicate the positions for lighting.

`RENDER`, and the `Render` dialog box appears **(Fig. 2.91)**. Make selections and the bracket is rendered as shown in Fig. 2.92.

2.26 Lights

A rendering is enhanced by point lights, distant lights, spotlights, and ambient lighting under your control.

Point lights emit rays in all directions from a point source.

Distant lights emit parallel beams like those of sunlight.

Spotlights emit a cone of light to a target surface.

Ambient light comes from no particular source and provides a constant illumination to all surfaces of an object.

New Lights

Light sources can be added to a drawing and indicated with the symbols shown in **Fig. 2.93**. From the `Render` toolbar, select `Lights` to get the

LIGHTS DIALOG BOX

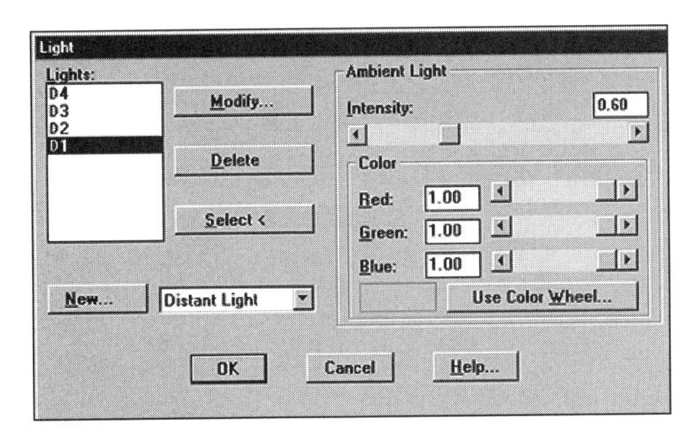

Figure 2.94 The `Lights` dialog box is used to set and modify lights for rendering.

NEW DISTANT LIGHT DIALOG BOX

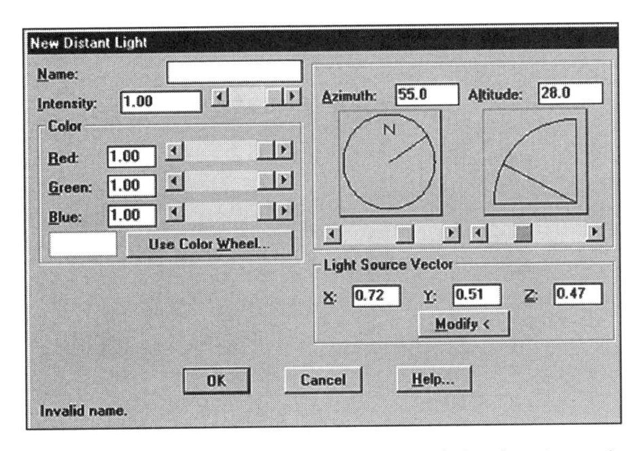

Figure 2.95 The `New Distant Light` dialog box is used to create and name a new light source.

`Lights` dialog box shown in **Fig. 2.94** where several lights are listed. Select `Distant Light` next to the `New` button, pick `New` to get the `New Distant Light` dialog box **(Fig. 2.95)**. From this box, you can name the distant light, set its directions, and pick its colors. Light `Intensity (0=off, 1=bright)` is set with the slider bar or by typing. Leave intensity set to `1`, select `OK` to return to `Lights` where `D2` is listed, and pick `OK` to exit.

NEW POINT LIGHT DIALOG BOX

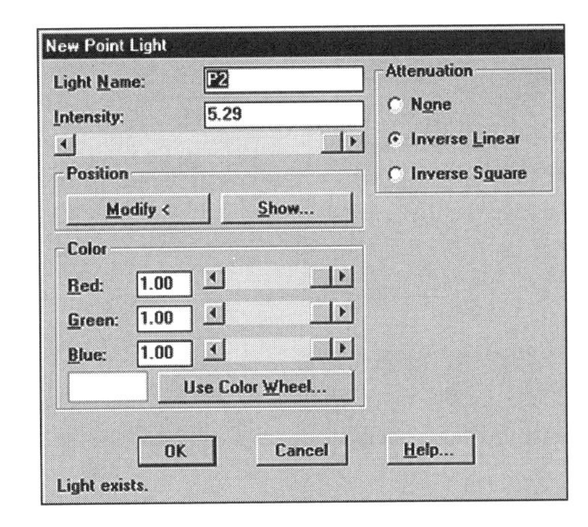

Figure 2.96 This menu is used to create a `New Point Light`.

MODIFY DISTANT LIGHT DIALOG BOX

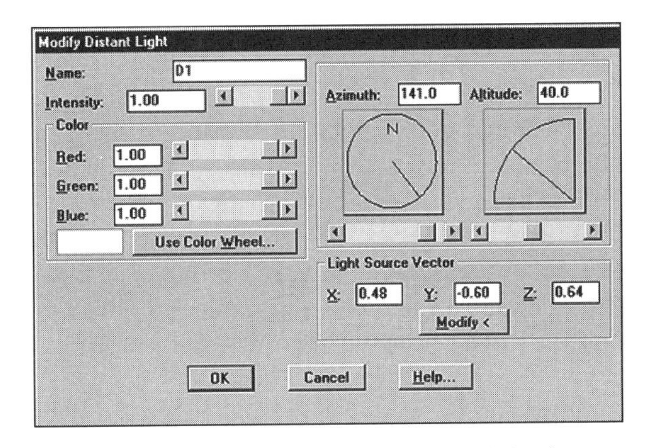

Figure 2.97 The `Modify Distant Light` dialog box is used to make changes in an existing light.

Select `Render` from the `Render` toolbar to obtain a new image of the bracket using distant light `D2`.

If you had selected `Point Light` and `New`, the `New Point Light` dialog box would have appeared on the screen **(Fig. 2.96)**. This dialog box is different from the `New Distant Light` box since point lights have different characteristics.

MODIFY COLOR

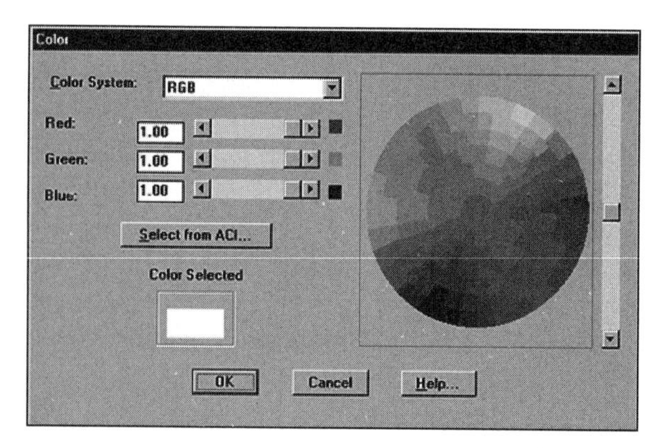

Figure 2.98 The Color Wheel is used to assign colors to lights and objects.

MOVING A LIGHT

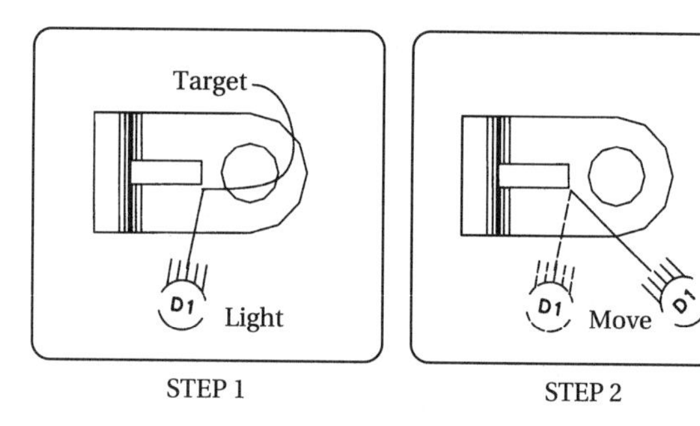

Figure 2.99 Moving a light:

Step 1 Select Modify from the Modify Distant Light dialog box (Fig. 2.97) and you are transferred to the drawing. Enter light target <current>: (ENTER) (Select current target)

Step 2 Enter light location <current>: .XY (ENTER) of (Need Z): 6 (ENTER) (The Modify Distant Light box reappears; Click on OK to exit and Render the bracket.)

Modifying Lights

Select Lights from the Render toolbar, select Distant Light to obtain the dialog box shown in Fig. 2.95, and double click on D1 (distant light) to obtain the Modify Distant Light box **(Fig. 2.97)**. The Use Color Wheel button can be selected to obtain the Color dialog box **(Fig. 2.98)**. Experiment with color intensity at other settings (for example, 0.2 and 0.6) and render an object to observe variations in the brightness of the image.

Moving Lights

To move distant light D1, type PLAN to get a top view of the bracket, where the light sources are shown as symbols. Each light can be moved with the MOVE command. You can also select Modify from the Lights box, pick distant light D1, and pick Modify to obtain the Modify Distant Light box **(Fig. 2.99)**. The Modify Distant Light dialog box appears (Fig. 2.97). Select Modify under the heading, Light Source Vector, and the screen returns to the drawing where you are prompted for the direction of the light rays. The direction can also be changed by the Azimuth and Altitude adjustments (Fig. 2.97).

RENDER TOOLBAR

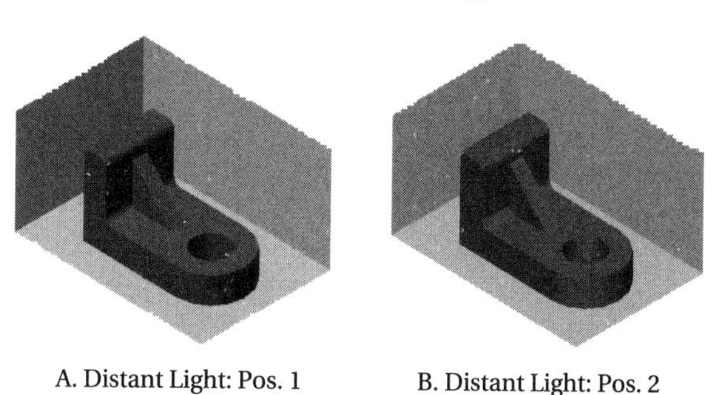

A. Distant Light: Pos. 1 B. Distant Light: Pos. 2

Figure 2.100 A comparison of rendered views of the bracket with Distant Light D1 in positions 1 and 2 is shown here.

In step 1, the prompt Enter light direction TO <current>: appears with a rubber-band line attached to the target. Press (ENTER) to retain the current target and obtain a rubber-band line attached to it and the prompt Enter light direction FROM <current>. In Step 2, type .XY, pick the light location with the cursor and respond to the prompt, of (Need Z) with 8 to

POINT LIGHT FALL-OFF (MODIFY LIGHTS)

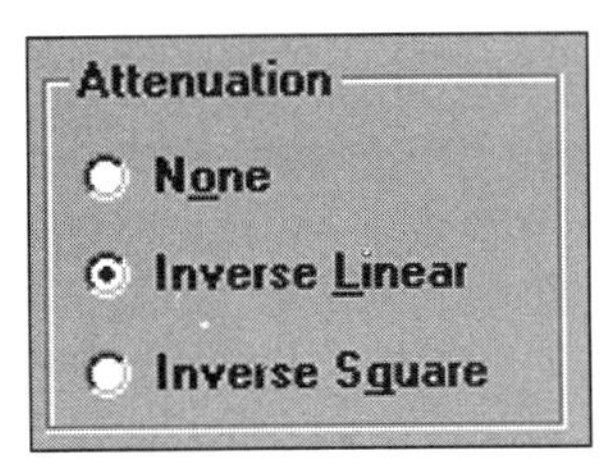

Only Point Lights have fall-off. Distant lights do not.

Figure 2.101 The type of point-light fall-off can be specified by selecting either Inverse Linear or Inverse Square lights.

locate the light source and establish the direction of the parallel light rays.

Type VPOINT and settings of 1,-1, 1 to obtain an isometric view of the bracket. Select Render from the Render toolbar to observe the original and new light position in **Fig. 2.100**.

Light Fall-off

The characteristic whereby light becomes dimmer as it travels farther from its source is called fall-off. Point Lights and Spotlights are affected by fall-off; Distant and Ambient Lights have uniform intensity at all distances. From the Lights box, select P1 (Point Light), select Modify to get the Modify Point Light box, which has an Attenuation area. The options available here are None, Inverse Linear, and Inverse Square **(Fig. 2.101)**:

None is a light with no fall-off, giving all objects the same brightness.

Inverse Linear makes a surface that is 4 units from the light one-fourth as bright, and one-eighth as bright when 8 units away.

Inverse Square makes an object 4 units from the light one-sixteenth as bright, and one-sixty-fourth as bright when 8 units away.

A comparison of lighting set to Inverse Linear and Inverse Square is shown in **Figure 2.102**.

RENDER TOOLBAR

A. Inverse Linear B. Inverse Square

Figure 2.102 Light Fall-off settings in the Light dialog box are None, Inverse Linear, and Inverse Square. By using Inverse Linear, an object 4 units away is one-fourth as bright as one 1 unit away from the light. By using Inverse Linear, an object 4 units away is one-sixteenth as bright as one 1 unit away.

AMBIENT LIGHT RESULTS

A. Ambient Light=0.3 B. Ambient Light=0.85

Figure 2.103

A Ambient light is set to 0.3.

B Ambient light is set to 0.85.

Ambient Light

So far, ambient light has been set to 0.3. From the Render toolbar obtain the Lights dialog box and set ambient intensity to 0.85 by typing or using the slider bar. The variation in lighting the bracket is shown in **Fig 2.103**.

Spotlights

Spotlights emit cones of light as defined in **Fig. 2.104** to highlight features. The creation of a

LIGHT SYMBOLS

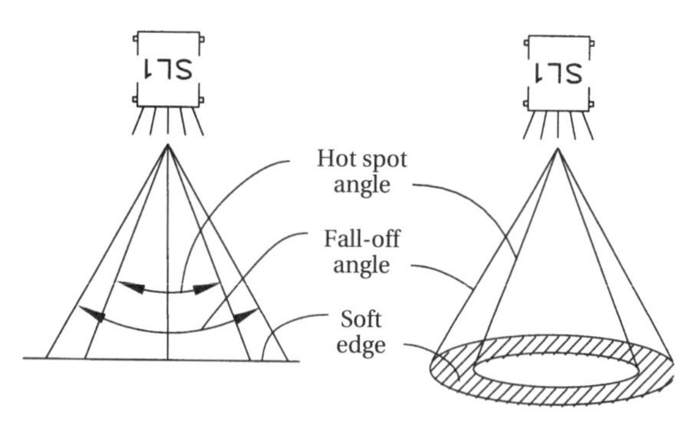

Figure 2.104 The definitions of the characteristics of a spotlight are shown here.

CREATE NEW SPOTLIGHT

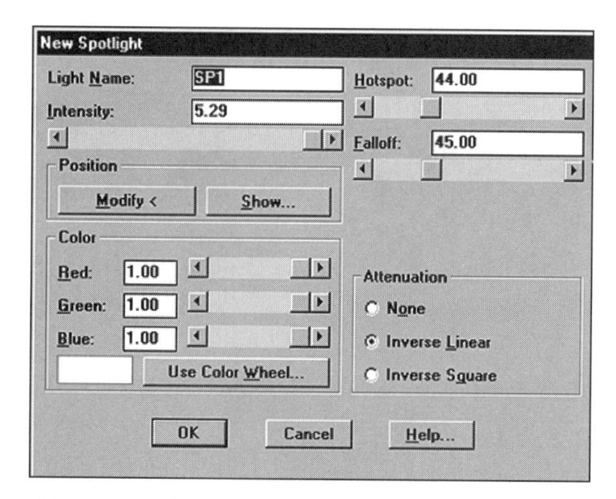

Figure 2.105 The `Create New Spotlight` box lets you create a light with a conical beam.

`Spotlight` begins from the `Lights` dialog box (Fig. 2.94) where `Spotlight` and `New` are selected to display the `New Spotlight` box in **Fig. 2.105**. You can see the effects of a spotlight in **Fig. 2.106**.

2.27 Working with Scenes

`Views` and light settings can be saved in combinations as `Scenes`, which can have several lights (or no lights), but only one `View`. A viewpoint of a drawing can be saved as a `View` by typing `VIEW` and giving it a name. Other viewpoints are saved in this manner. To recall a `View`, type `VIEW` and `RESTORE`, and give its name when prompted.

A SPOTLIGHT APPLICATION

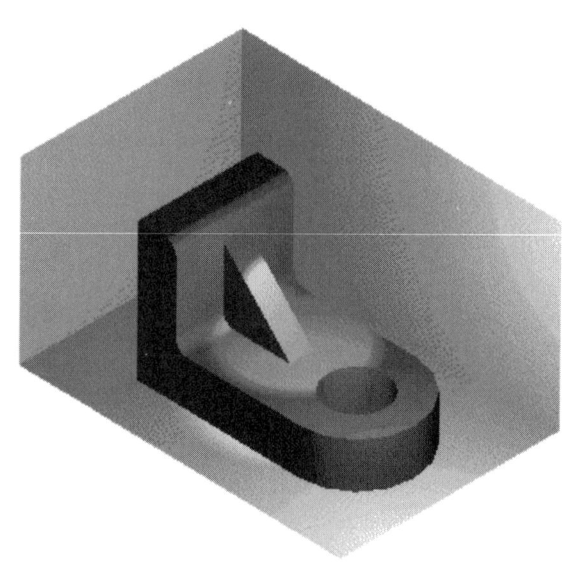

Figure 2.106 An example of a spotlight applied to the bracket.

Restored `Views` of the bracket used in the previous examples are selected from `Scenes` of the `Render` toolbar to obtain the `Scenes` dialog box **(Fig. 2.107)**. Select `New` and the `New Scene` box appears in which all model-space `Views` and lights are listed. Type the name of the `Scene` in the box (`S2`, for example). The name is truncated to eight characters if it is longer than eight. Select `V2` as the `View` in `S2`, select pointlight `P1` and distant-light `D1`, select `OK`, and the `Scenes` dialog box reappears. Other scenes can be created by using these steps.

If `*NONE*` were selected as the scene to render, all lights would be on in the current view. If no lights existed, an over-the-shoulder distant-light would be given.

SCENES DIALOG BOX

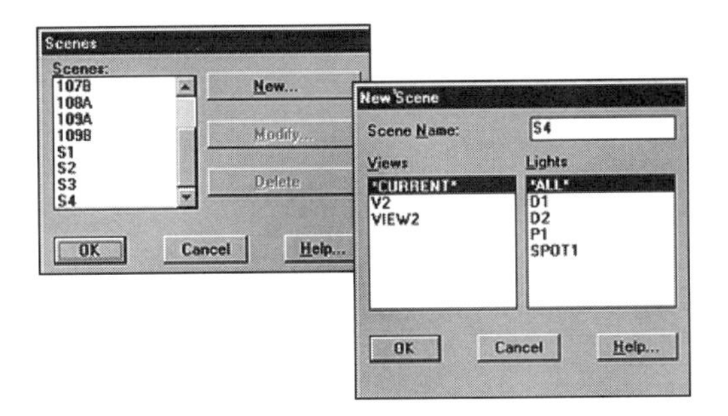

Figure 2.107 The `Scenes` dialog boxes are found under the `Render` toolbar. They are used to select lights and views that comprise scenes.

SCENES

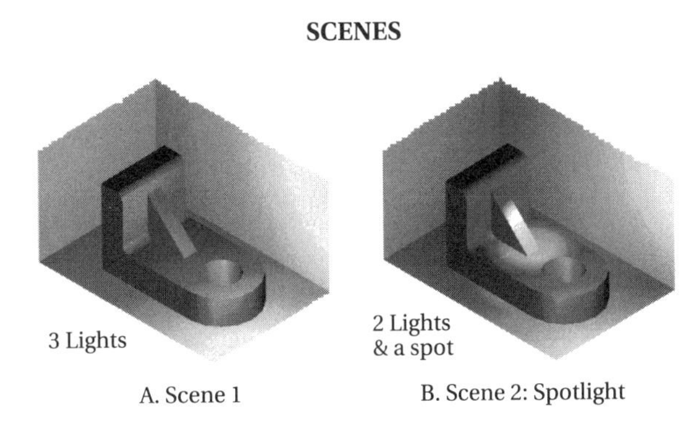

Figure 2.108 Examples of scenes are shown here with different combinations of lights.

Select `S2` from the dialog box, pick `OK`, and choose `Render` to obtain a rendered `S2`. A comparison of `S1` and `S2` is given in **Fig. 2.108** after they have been rendered.

A `Scene` can be modified by selecting it from the `Scenes` dialog box (Fig. 2.107) and picking `Modify` to obtain the `Modify Scene` dialog box, which is similar to the `New Scene` box. Different lights can be selected or deselected. Select `OK` to keep your modifications and return to the `Scenes` dialog box, select `OK` to exit and `RENDER` the modified `Scene`.

MATERIALS LIBRARY DIALOG BOX

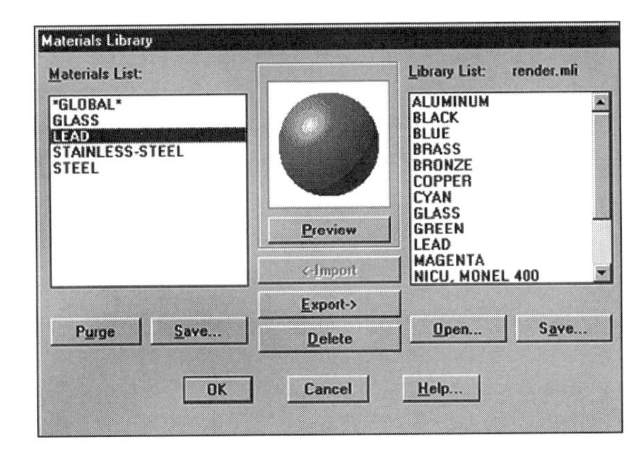

Figure 2.109 `Materials Library` dialog box lists the materials and colors that can be selected and assigned to parts. A preview of their applications is shown on the sphere.

2.28 Materials

The material of an object determines the reflective quality of its surfaces from dull to shiny. Select the `Materials Library` icon from the `Render` toolbar to get the `Materials Library` dialog box **(Fig. 2.109)**. Materials can be selected one at a time from the listing and previewed by selecting the `Preview` button. To add a material to the `Materials List` for future use, pick `<Import`. Click on `Save` to add the material to the `Materials List` file. Select `OK` to exit the `Materials Library` dialog box and return to the drawing.

From the `Render` toolbar, select `Materials` to get the dialog box shown in **Fig. 2.110**. To `Attach` a material finish to a part, select it from the list, select `Attach`, and the drawing reappears. You are prompted, `Select objects to attach "MATL" to:` Select the part and the `Materials` dialog box returns; click on `OK` to exit. `RENDER` the object to see the results.

MATERIALS DIALOG BOX

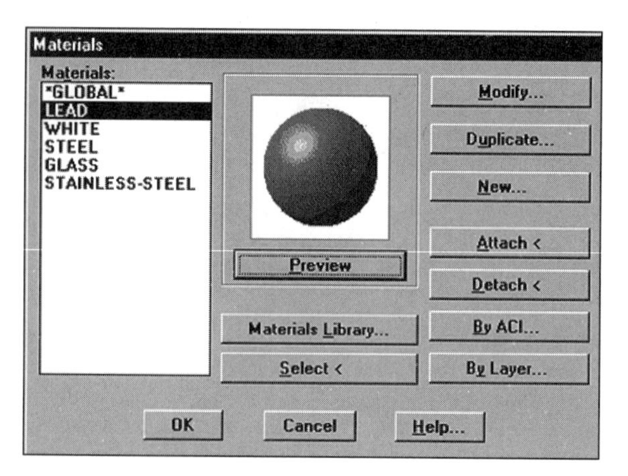

Figure 2.110 The `Materials` dialog box is used to preview various materials as they are assigned to a part.

MODIFY STANDARD MATERIALS

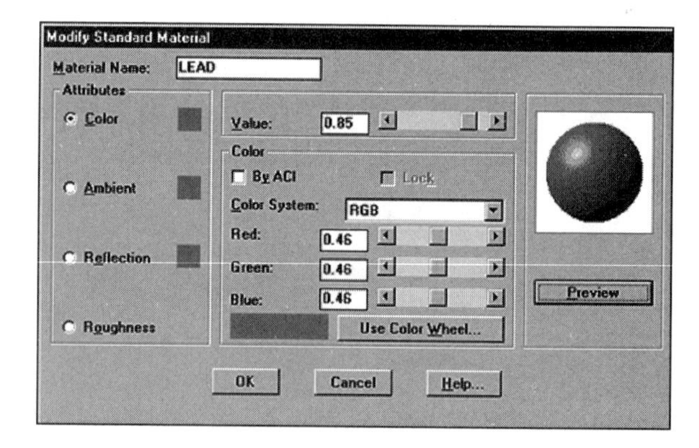

Figure 2.111 Selected materials can be modified by using this menu.

By selecting `Modify` from the `Materials` dialog box (Fig. 2.110), the `Modify Standard Materials` box **(Fig. 2.111)** lets you modify the selected material. These settings are `Color`, `Ambient`, `Reflection`, and `Roughness`. Experiment with different settings and observe the results of these changes by picking `Preview`.

Index